ENERGY BASIS FOR MAN AND NATURE

ENERGY BASIS FOR MAN AND NATURE

Howard T. Odum

Environmental Engineering Sciences
and Center for Wetlands
University of Florida
Gainesville, Florida

Elisabeth C. Odum

Santa Fe Community College
Gainesville, Florida

McGRAW-HILL BOOK COMPANY

New York   St. Louis   San Francisco
Auckland   Düsseldorf   Johannesburg
Kuala Lumpur   London   Mexico
Montreal   New Delhi   Panama
Paris   São Paulo   Singapore
Sydney   Tokyo   Toronto

ENERGY BASIS FOR MAN AND NATURE

1234567890   KPKP   78321098765

This book was set in Times Roman by Black Dot, Inc.
The editors were B. J. Clark and Susan Gamer; the cover was designed by Joseph Gillians; the production supervisor was Charles Hess. The drawings were done by J & R Services, Inc. Kingsport Press, Inc., was printer and binder.

Alternative Sources of Energy, from which several drawings have been reproduced, is a quarterly published by Don Marier, Route 2, Box 90-A, Milaca, MN 56353.

The drawing on the opposite page was done by Ann Odum.

Library of Congress Cataloging in Publication Data

Odum, Howard T   date
    Energy basis for man and nature.

    Includes bibliographical references and index.
1. Power resources.   2. Economics   3. Environmental policy.   4. Human ecology.
I. Odum, Elisabeth C., joint author. II. Title.
TJ163.2.038   333.7   75-23249
ISBN 0-07-047531-8
ISBN 0-07-047527-X pbk.

Contents

PART TWO   ENERGY SYSTEMS SUPPORT HUMANITY

PART THREE   ENERGY CRISIS AND BEYOND

Preface

This book is aimed at the general public, political leaders, and students who may need a short concise statement of the principles of energy and the way these shape our culture, past, present, and future. Understanding energy is critical during times of changing energy conditions. There may now be mistakes in government policies on energy and value; and there may in the future be fundamental changes in the kind of life that is possible. Before 1973, energy was expanding; since then, it has been leveling off or declining. This book introduces simple, commonsense diagrams that use the energy principles to visualize broadly the patterns that affect our own nation and the world. We hope these overviews will provide a general way for all of us to understand the place of humanity in nature, the responses of our system, and a way to predict the future.

This book includes patterns of systems, the basic laws of energy, alternatives open to human beings, and a glimpse at some possibilities for the future. A language of simple, visual symbols is used because we have found it successful in our teaching of energy and systems. The concepts and ideas we present, therefore, are expressed both in words and in diagrams. More elaborate discussions of the topics in this book are available in a previous book, *Environment, Power, and Society.**

Part One of this book introduces energy principles and the flows of energy in our environment. Part Two discusses the energy basis for humanity. Part Three examines possibilities for the future. Following Part Three are a glossary, a summary of symbols used, the mathematical formulas for the models in Chapter 5, suggested exercises and activities for each chapter, and a list of references. The exercises and activities are provided for those in formal teaching situations and have been tested in classrooms in several schools.

ACKNOWLEDGMENTS

We acknowledge the extraordinary intellectual ferment on energy and environment at Gainesville and especially the contributions from students and faculty that formed the background for this book. Our work was encouraged by an atmosphere of open inquiry generated by the administration of the University of Florida. Groups and projects involved with energy included: Systems Ecology, Environmental Engineering, Center for Wetlands, Carrying Capacity Studios in Architecture, Life Sciences, Energy Center, Forestry and Wildlife Management Unit, Center for Latin American Studies, Sea Grant Program, and projects of the Rockefeller Foundation, the Atomic Energy Commission, the Energy Research and Development Administration, the National Science Foundation's division of Research Applied to National Needs, U. S. Department of the Interior, the Florida Division of State Planning, the Florida Division of Pollution Control, and the Department of Transportation. Professor Chester Kylstra collaborated in directing our net-energy research project.

We are grateful for the innovative policy of Santa Fe Community College, which encouraged the experimental testing of a preliminary version of this book in a new energy course.

Howard T. Odum
Elisabeth C. Odum

*By H. T. Odum, John Wiley, New York, 1971.

ENERGY BASIS FOR MAN AND NATURE

Introduction

Everything is based on energy. Energy is the source and control of all things, all value, and all the actions of human beings and nature. This simple truth, long known to scientists and engineers, has generally been omitted from most education in this century. When energy sources are rich, economies, knowledge, and aspirations grow; when energy sources are all being used as fast as the earth receives them, activities, values, and aspirations settle into a steady pattern. So it has been throughout the history of humanity and nature.

Only when sources of energy are newly available and rich do people feel free to do what they want as individuals. The freedom to make many choices exists during only a brief period. After that, competition among people and methods limits choices to those that make use of energy most effectively. Americans, along with the rest of the world, are coming out of a century when there was an excess of energy and much freedom of choice was possible. We are moving into a period characterized by less energy.

A steady period differs from a period of growth in many values and ways of doing things. The transition from growth to a steady state can be smooth and planned so that individuals can make changes and learn new ways; or it can be sharp, disorderly, and disastrous for individuals, with unemployment, famine, epidemic diseases, and unnecessary war.

This book discusses how energy controls our lives, our economy, our international relationships, our standard of living, and our culture. To understand this story, one must have an understanding of humanity as a part of nature. Second, one must acquire a modern systems view; for this purpose, we will use a language of simple systems diagrams which explain trends such as galloping inflation in terms of their underlying cause: the shifting basis of energy. Third, one must shake off preconceived ideas of what trends exist and what is good.

The options available for the future are set out for us by the laws of energy. As patterns of energy change, so do human roles. When national governments are confused about the future, the reason is that elected officials, journalists, and specialists fail to understand the primary role of energy and are making decisions about problems without knowing that energy is the cause of the problems they are dealing with. The inexorable march of events will ultimately force all people to understand what must be, but this may come so late that transitions can no longer be made in an orderly way and with protection for individuals. Those cultures that can adapt will prevail. If enough members of our culture can be flexible of mind and enthusiastic about this chance to lead, we can make a safe transition to a steadier state.

FOR WHOM THE BELL TOLLS

When we call for leaders, we speak to all, but with a different emphasis for different readers.

The uncommitted student may have to adopt a more realistic viewpoint and question some of his previous education regarding the rights of individuals by asking the question: Does this activity use energy effectively? That is, is it *energy-effective?*

The teacher must make the laws of energy of primary importance at all levels of teaching: the energy laws are the driving principles of the world. In every field, in every activity, and in every lesson, energy-effectiveness should be gentle on the mind. Energy is not only for specialized courses in physics and chemistry; it should permeate education from kindergarten through graduate school. And specific courses in energy are also needed at various levels.

University administrators must ask if their curricula address the real process of fitting with nature or whether they simply foster the status quo. In the future, resources for maintaining and teaching knowledge will be ever scarcer. Because we will be able to hold and use less information, sharp choices must now be made. The administrator must decide what is essential and how it can be retained for less cost in terms of energy.

The citizen must ask that government show how effective each proposition is in terms of energy. This must include its energy cost. Citizens must no longer accept money as the criterion of value. They must no longer leave energy to specialists who do not worry about the effects of energy on the economy as a whole.

Citizens who think of energy as simply one commodity, separate from matter, information, art, and human spirit, must learn that everything has an energy component. The more intangible and valuable something is the more it costs in energy. And the more intangible a value is the more energy value is lost when it deteriorates or is lost. Those who think that energy determinism is a vague, unproven theory must dig deep, learn, and watch. Every day in the newspapers we can read about the effects of changes in energy and watch predictions about energy unfold as the great growth of humanity changes to long-range patterns. These changes may bring us nearer the goal of having steady hands on the cycles of the biosphere.

The politician must interact with his constituency so that his proposed plans, laws, and programs reflect the new need for evaluations in terms of energy as the basis for all choices. He must help his constituents to learn as he learns, so that the beliefs they have acquired during several generations of living in a period of excess energy can be corrected in the face of a new reality. This is an opportunity for new politicians to overthrow incumbents who are slow to learn. But it is also an opportunity for incumbents to learn along with the voters and thus strengthen their position.

The environmentalist must learn to measure the value of the environment, and of proposals concerning it, in terms of energy-effectiveness. Environmental protection must have as its goal a symbiosis of humanity and nature rather than a competition in which funds are squandered and opportunities for redesigning are wasted.

Members of governments who prescribe absurd energy costs in the name of environmental protection—such as cooling towers, emission devices on cars in rural regions, and tertiary sewage treatment—must acquire a better view of the symbiotic relationship between humanity and nature. Many devices are not needed, because some heat and wastes are a resource to the biosphere if they are not too concentrated; the basic problem is too high a density. Criteria must be adjusted for survival in a world where economies compete for energy; prepare plans for declining densities of urban activity.

Physicists, who may think themselves authorities on energy, must change their emphasis. They must learn about energy in the ecological system. The synthetic aspect of energy is a different story from the reductionist aspect. The nuclear physicist must consider the possibility that very-high-energy phenomena like hydrogen bombs, although natural on a larger scale of the universe, may be too intense to yield net energy under conditions on earth. Much nuclear energy must be thrown away in unproductive work simply to bring the temperature down to a usable range.

Biologists who have used energy for budget-sheet accounting in biological processes must learn that the bomb calorimeter does not give the energy value of a high-quality substance. The "energy value" of a person is much more than the heat obtained when he is burned. Nutritionists must learn that 5 Calories per gram is not the energy value of protein when protein is used at its maximum value in building living tissue.

Geologists must stop overestimating the reserves of fossil fuel in the earth, and must stop using the same value for surface oil seepage as for oil that is 4 miles deep and 20 miles out to sea. Geologists must acquire the concept of net energy and consider the high-energy feedbacks in the steel construction of oil rigs and the training of engineers.

"Basic scientists," who define *basic* as "looking to the parts," need to learn that putting parts together to understand whole systems is equally basic. The scientist who says that synthesis is "applied," as if it were an inferior activity, must ask which is intellectually more difficult and ultimately more basic, reductionism or synthesism. Surely both are necessary, but we have had too little synthesis, and our science curricula in schools have failed to fulfill their promise because of this.

The scientist who uses his discipline to learn more and more about less and less must connect his specialty to the real world as an entirety. Anyone who sets boundaries to his field of interest is limiting his capacity to grow. An old discipline has already yielded what it can; now knowledge must be arranged in different ways and given different names. New disciplines need to be examined for what they can contribute.

Engineers must use their techniques for measuring and evaluating energy systems on a larger scale, considering both man and nature and including human economy in their calculations. Engineers should realize that most technological advances during the last century of growth have involved the application of hidden, indirect, additional forms of energy. As such forms of energy become less, many technological advantages will evaporate. What was an advance becomes wasteful and must be discarded. Efficiencies must be calculated as the ratio of outputs to *all* energy inputs, including the ultimate energy value of bought goods and services. Engineers have to recognize that energy has different qualities. Different kinds of energy must be converted to a common base, with quality factors that indicate their real worth in support of the economy of man and nature.

Activists who see individual rights as the goal of political and social action must learn that these ideals are only partly correct as far as survival of their culture in a world of declining energy is concerned. They must understand that the individual has the right to form a symbiotic relationship with his system, but that this right demands a meaningful contribution from the individual. Freedom of choice is necessary for the individual to find the best relationship, but this freedom is less for the sake of the individual than for the sake of survival of both individual and system.

The feminist and antifeminist must recognize that in a steadier, more level state, if some of our present level of medicine and regulation of disease is retained, a high rate of reproduction is not needed. The full use of women in the work force is essential to the vitality of our economy and culture. Whatever changes in custom and law are necessary must be made. The feminist must learn that these will have as their purpose not the rights of the individual per se, but the right of the individual to better serve the large system of humanity and nature. Real differences in sexual characteristics must be recognized whenever they have effects on service which are energy-effective.

Leaders of minority groups must stress the opportunity for the system to become more effective by fully utilizing the potential of those who are now inadequately educated and poorly motivated. Talk less about rights for the sake of individuals.

The lawyer must learn that in times of lower energy many processes of justice and regulation will take place at the level of individuals and small groups rather than at the level of laws and institutions. More law must recognize energy-effectiveness as a principle. Lawyers must deal less in minutiae and more in helping people understand legal teachings and laws. Some legal functions may be taken over by religious bodies and individual codes. The laws that went along with growth may be poorly adapted for a period of leveling off. In some cases it may be well to wipe out whole codes and replace them with simpler versions based more on energy principles.

Police forces must plan for smaller-scale law enforcement which uses less energy. Populations will become less mobile, and local social organization stronger. The police should plan for fewer cars and computers. The work load per person may increase as resources decline. During the period of transition, disorder may increase temporarily.

The economist must learn how energy sources work. He calls such sources *externals* and often is unaware that their flows control the economy and cannot be ignored. Economists have recently recommended stimulation of growth by means of money manipulations. However, this works only when there are large, unused energy resources. Now productivity is declining as energies decline, and manipulating the circulation of money will have little effect. Economists must learn that energy laws are primary.

Those who maintain that higher prices will end any shortage must learn that when the cost of processing energy resources rises, a larger percentage of the total economic effort must be used for this processing. As a result, less energy is available for other uses and there is a general decline in what we can do. When the cost of finding and processing energy sources rises, the standard of living and the energy used on behalf of each person will decline. Ultimately, prices are determined by energy.

Leaders in business and banking must recognize that large-scale capitalism, where money is always able to generate more money, exists only when energy is expanding; it disappears when a steady state is reached. Capitalism will continue temporarily in a few countries, but not in those with declining energy sources. Loans, borrowing, banks, stocks, bonds, etc., will be less and on a smaller scale. Many people now involved in those activities will transfer to others. The identification of capitalism with Western culture will decrease.

Wealthy people who still have extra money for starting new endeavors must show the way. They should stop backing growth, stop serving individual pleasure, and realize that their money is very, very temporary. It should be used for planning for a steady state. In a declining-energy system money does not earn money; it inflates away. The rich should prepare to join those of average means.

Optimists in the chamber of commerce, who think we "grow or die," must learn about steady-state ecosystems which have worked evenly and cleanly for 100 million years. They must realize that the role of growth is often to prepare a system for a steady state.

Manufacturers of goods and managers of services that accompany frivolous activities—such as driving big cars and indulging in excessive tourism—must make early adjustments. They must reduce the energy levels of their activities, reduce staffs, and plan a gradual transition to smaller units, lower costs, and more emphasis on essentials. A temporary exception is that production which is part of economies of oil-rich nations.

Managers of large-scale distribution—the supermarkets and the chain stores—must expect a declining scale. Local competitors will come back. The shopping plaza may change into a local center rather than a regional center, as energy for transportation becomes scarcer.

Those who work in transportation must recognize that the need for transportation will decline. When there is a choice as to what to repair, the criteria should be energy-effectiveness for the society as a whole. Speed must drop and with it many of the designs that produce speed but are costly in terms of energy.

Managers of airlines must recognize that the extremely high energy costs of air travel will limit flying. Speed will be justified only where the energy-effectiveness of a trip is high. This implies fewer trips, smaller and slower planes, and much less international travel except to and from places where energy is still growing.

People in the wood and paper industries must understand that low-energy systems are unable to maintain the present level of information processing. Wood will be used more for basic building and less for paper; some forest lands will be shifted to food agriculture. There will be less mechanized management of forests and more labor-intensive management and self-maintenance of ecosystems.

Advertising and public relations people must recognize that there will be less need for information to spur growth. Instead, they should help educate the public for the transition to the steady state. Some of them will have to find new work; others must learn to appeal to the new goals of the steady-state system. All must learn that they are not free to mold public opinion as they see fit. The reality of limited energy will determine what is acceptable.

The land developer must see that the value of land in terms of urban development and proximity of tourist attractions will give way to the value of rural land. The value of land used for agriculture based on solar energy, for forestry, and for low-energy tourism will increase. Land will be needed for uses that are labor-intensive.

The mayor with urban problems must plan for the best use of existing buildings in the central city for housing a small amount of business without energy-expensive urban renewal. The cities may be restructured into small centers, with the central cities again housing people but at a lower density. Costly activities which are now done by machine but could be done by hand—such as garbage collection—will again be done by hand. This may require categories of lower pay for young apprentices and the elderly.

Architects must abandon their view of themselves as artists doing what fancy will without real attention to the energy-intensiveness of the function. Plans for good use of old buildings will become more important. Repair will supersede replacement. Buildings and landscape plans will have to use enough natural space so that replacement of parts need not involve renewal of entire districts. Zoning must be discarded so that diversity in a steady state can replace the homogeneity of "grow and die" districts. Air conditioning will generally be discarded. The era of "artistry in concrete" is almost over.

Union members will lose their power to advance wages when what is required is a cut in everyone's pay. This still leaves the unions a large role in helping individuals adapt to the new system. The unemployed will need advice and assistance in getting new jobs. In lower-energy work like agriculture there will be less need for machines and more need for people.

People in agriculture must recognize that the energy subsidy available to agriculture is declining, reversing a two-century trend. More land will be required; more labor will be required; more people will turn from suburbia to the farms instead of to the cities. Agribusiness will be replaced by more "agrihumanity." The use of roles natural to the ecosystem, such as the control of insects by birds, will increase. Rotation of crops will replace chemicals and intensive machinery. Unemployment in the growth and luxury industries will ultimately shift people to agriculture. Plans for this transition are needed. Agricultural products will be sold over a smaller area. There will be less international exchange, less regional specialization, and more diversity on a small scale instead of diversity on a large scale. Home extension agents who have forgotten how to farm without poison must go back to school, as soon as the agricultural schools put courses in lower-energy farming back into the curriculum.

SOME BASIC PRINCIPLES

Forgive us if we claim too much in the foregoing paragraphs. Our aim has been to encourage our readers to master the energy diagrams we will present and the principles involved in them. But the reader should not believe too little and thus miss a chance to understand the new system. It will do little good if you admit five years too late that the laws of energy could have shown you the correct way: by then your chance to serve may have passed.

This book starts with energy principles, shows how systems of human beings and nature are determined and change, draws illustrations from ecological systems and human history, and finally suggests how the future is predicted from basic energy principles. These methods correctly predicted the recent economic decline and can help us to foresee the future.

Before we turn to the principles of energy and the systems in detail, let us briefly discuss some of the main ideas. Experience in lectures, press conferences, and teaching has taught the authors that most Americans do not at first follow, understand, or accept the reality of the control of their system by energy. To understand reality, the diagrams are necessary; but people will not go to the trouble to grasp the diagrams unless they are somehow convinced that this is important. We hope that this introduction will suggest the story of energy before we go on to its details. Once the story has been sketched in, we will let it emerge again stepwise, factually, using some of the language of science, engineering, ecology, and economics.

Three energy laws will be presented: the conservation of energy, the necessary degradation of energy, and the maximization of effectiveness in the use of available energy sources. From these laws we learn a common pattern of transformation of energy. This pattern produces some energy with increased quality, stores it as assets, and then uses its special qualities as feedback to help draw in more energy. The kind of energy available in a region determines the characteristics and activities of that region. Energy resources control the actions of nature and humanity in an area. Depending on these energy resources, at first power is maximized by growth, then power is maximized by characteristic transitions, and finally energy declines and a steady state is reached. The graph of the movement from growth to decline to steady state is determined by the pattern of energy resources. We can forecast it using the language of symbols which this book presents. The time graph of the action of energy resources may be found by computer, using mathematical equations equivalent to the diagrams. See page 271.

When there is an unused flow or storage of energy, assets are fed back to accelerate growth. When the flow of energy is renewable and steady, the system builds a structure that can be maintained in a steady state. When a system starts with some renewable steady energy and some unrenewable storage like oil, there is a burst of growth, then a decline in the level of energy, and finally a leveling off to the steady state.

When the language of energy is used, all kinds of systems can be examined comparatively, including ecosystems, economic systems, geological systems, meteorological systems, and cities. Similar patterns of use of energy occur in all these systems. A chain of energy-transforming units builds order and increases the quality of part of the energy. The high-quality energy feeds back to make possible more processing of energy.

Money circulates as a countercurrent in only part of the system. The ultimate vitality of an economy depends on the effective interaction of all the free and bought energies of man and nature. The best pattern is the most energy-effective, in which there is maximum use of resources for survival.

Many corollaries and principles emerge from the consideration of laws of energy and interaction of systems. By tracing the development of the ecological systems and then the increasing role of man in them, we see the reason for many changes that have not always been well explained in typical historical accounts. We gain many insights into war, disease, old age, and what should be kept immortal. Patterns of succession, leveling, decline, and steady state are predicted from the models. We see reasons for much of the energy and money crises actions that are affecting the world in the 1970s. From the large-scale patterns we can see and speculate about how these are affecting our individual lives. Ultimately, the individual can look ahead to the future. This book can help the reader to see what the majority will later believe, and to prepare himself for changes.

Our problems with energy, inflation, and the economy, and the changes to come, become startlingly clear. It turns out that the net energy from many sources is negative or small because more hidden energy inflows than we realized. Some forms of energy are of low quality and are too dilute to do much until they are concentrated at a great cost of energy. Ordinary wind and waves, for example, and ordinary geothermal gradients yield little net energy.

Solar energy, the great renewable source of the biosphere, is usable only after it is concentrated. Various schemes for harnessing solar energy turn out to be installations based mainly on fossil fuels, with their main energy flows not really supported by the sun. On the other hand, we will find that the long-term basis of our economy is ultimately the use of effective self-organizing solar converters: forests, ecosystems, and lower-energy agricultural patterns that have long been with us.

Nuclear energy does not yet yield net energy. Even if the present plants last as long as they are supposed to, without major accidents or deterioration, they will yield less energy per unit of energy invested than other sources. Another principle of energy applies here: it is impossible to build new structures when there has been no recent growth to generate the necessary capital. This essentially eliminates such alternatives for gigantic expansions of energy as atomic energy. Atomic fusion is still uncertain; but consideration of net energy, capital cost, and possible complexity indicates that it may be even less energy-effective than the present nuclear stations which use the fission process. Our technology may be unable to use extremely intense energy such as fusion (as in hydrogen bombs) because too much energy is required to maintain control as it is diluted to the intensity of the human system.

In short, many policies proposed during national and international discussions to augment energy and continue to grow are incorrect and will not do what their proponents hope. We must accept the reality of declining energy and plan for it. The most urgent need is for national and international task forces to create an orderly transition to a state of lower energy followed by a steady state. The burden of proof is on those who suggest ways of finding new net energy, for our calculations make us skeptical about such proposals.

We must also challenge the idea that in a state of declining energy one can do more with less. Maintaining complex information involves a high cost in energy. Special programs are needed to preserve information as universities, libraries, and government information centers lose part of their budgets.

International exchanges of energy determine not only the balance of money payments but the level of energy support of individuals. The energy diagrams reveal that it may be a mistake to seek independence from foreign energy. An ultimate exchange of labor for machines in many basic production activities, such as agriculture and housing maintenance, is predicted. Unemployment will be a problem primarily for those changing over from industrial, business, and informational sectors that must decrease.

From the concepts of energy, a carrying-capacity formula is calculated. The amount of free energy one economy has determines the amount of additional outside energy it may purchase and still be competitive with other economies. The more energy your land area provides, the more energy it can draw in as investment. Ultimately the carrying capacity is determined by the environmental energy that you use well and the amount of special outside energies available to be purchased as matching energy.

Thus, energy gives us a way of projecting and planning the future, determining what level of human life best fits nature to create a vital economy, and making difficult choices for the ultimate public good. The brief statements of principles we have just made can probably be understood only in terms of the diagrams that will be presented in the text. These diagrams are necessary for a clear understanding of net energy, feedback, efficiency, yield, and the interaction of money and energy. We turn now to Part One, which has chapters on the fundamentals of energy and energy-effectiveness.

Part One

Flows of Energy Build and Operate Systems

Part One introduces concepts of energy and discusses various kinds of energy and measures of energy. Energy is the primary, most universal measure of all kinds of work by human beings and nature. Basic laws of energy apparently apply to all human processes and all processes of nature, including economics, culture, and aesthetics.

To help us visualize laws and flows of energy, some simple diagrams using energy symbols are introduced. These symbols should be learned at the outset, so that the rest of the text will be easy to understand. Along with the diagrams, we introduce the idea of a system as a combination of interacting parts. The energy diagrams are systems diagrams because they show how parts interact to produce overall effects. Using the diagrams, we illustrate principles of energy and show how flows of energy build systems, generate order, maintain themselves, circulate materials and money, and operate the world we live in.

Chapter 1 introduces the energy symbols and shows how they are combined to form systems diagrams. Chapter 2 defines energy. Chapter 3 gives the principles of energy. Chapter 4 describes the relationship of money to energy. Chapter 5 shows that patterns of growth and leveling occur when various kinds of energy are available. Chapter 6 indicates when choices for the use of net energy favor growth, exchange, and diversity. Chapter 7 gives examples of the flow of energy in ecological systems, in order to show some properties that may be generalized to apply to the changing patterns of human life. Chapter 8 presents the energy-driven cycles of the earth and the kinds of energy resources available on earth.

Chapter 1

Systems of Energy Flows

Faced with shortages of energy, galloping inflation, overgrowth, and concern for protecting the environment, human beings are coming to realize that they may be forced to change their patterns of life. It is now clear that our future depends on the connection of energy, economics, and environment (sometimes called *the three E's*) into one system of interdependent actions.

As piecemeal remedies and programs fail, many legislators and educators call for some better general understanding of the basis of life. General explanations of, and introductions to, the three E's are needed; courses in schools should be planned to reorient education toward such understanding. A commonsense overview of humanity and nature will show how the laws of energy control the human patterns, economics, times of growth, and times of leveling. Is the much discussed "steady state" a utopian ideal? Is it a state of economic ruin to be feared? Or is it an inevitable future to be welcomed?

This text looks at the world as a whole, considering how human beings fit into their environment. When it is understood that energy causes and maintains the order of man and nature, intelligent choices can be made for economic and political actions, and individuals can choose how to live in a world they understand. The key to understanding so much complexity is a simple concept: that of the energy flow. The modern concepts of energy discussed in this text are actually rooted in age-old ideas.

In all cultures, there seem to be concepts of energy and of man and his environment. Ideas of energy, survival, and systems permeate our own culture both in literature, and in science. Engineers, for example, measure energy quantitatively. Theologians may relate the idea of orderly development to work, which is the result of flows of energy; here is a quotation from the third chapter of Ecclesiastes:

> To everything there is a season and a time to every purpose under the heaven. . . .
> A time to be born and a time to die; a time to plant and a time to pluck up that which is planted; . . . a time to break down and a time to build up. . . . Wherefore perceive that there is nothing better than a man should rejoice in his own works for that is his portion, for who shall bring him to see what shall be after him.

As long as abundant energy was available for the rapid expansion of productivity and the growth of human culture, food supplies, technology, and knowledge—and, apparently, of the ability to survive—people were content to take energy, economics, and the environment for granted. Even the schools have emphasized other matters. Students considering their futures tended to think more about the inequities among people and about preventing harmful effects of great flows of energy and wealth—war, pollution, overdevelopment.

Cartoon by Frank Miller. Copyright 1974 Des Moines Register & Tribune Co.

Recently, however, the rapid growth which has characterized the past two centuries has begun to be restricted by limitations of supplies of energy. The energy situation was dramatized in 1973, when the Arab countries withheld oil from the world for a time. Much growth stopped, and attention was drawn to energy resources and how they interact with economics and the environment. At the same time, more rapid inflation began to reduce the buying power of individuals.

Many fundamental questions are being raised; many changes are taking place that are poorly understood; and confusion is resulting when opposed recommendations are put forward for solving the "energy crisis." Is growth to continue, or must we expect no further expansion? Will levels of energy decline, and with them the standard of living, the population, and the progress of the past two centuries? Are there other sources of energy, rich enough to replace those that are running out? What measures exist to indicate what is important in terms of energy? What national policies for money, prices, and loans will make for the best future? What policies should citizens support to ensure a stable future? What life-style should we expect in the future? How can the individual visualize and understand what is happening?

It is clear that most Americans have not had much education in basic laws of energy, in economics, or in principles of environmental interaction (sometimes called *ecology*). Much of our education has been aimed at examining *parts:* that is, in taking systems of man and nature apart for analysis. There are, of course, new concepts of putting parts together in order to understand the behavior of the whole; but few have the training to perform such synthesis.

One way of visualizing many parts in order to make complexity simple is called the "systems" approach. In this approach, diagrams are used to visualize systems, and from the diagrams calculations are made about flows and storages. For example, a plan for the water pipes in a new house is a systems diagram. From the plan we may learn how fast water can flow in and out, what maintenance will cost, and what kinds of energy are involved. Since energy is involved in all processes and events, diagrams can be drawn for all of them, from flows of water and growth of plants to world sales of wheat and international politics. Simple diagrams of energy flows help us to visualize how resources control what happens and to understand and predict the future.

Our first step, therefore, is to introduce the systems diagram, with some of its main symbols. Then, the diagrams will be used to show how the main parts and processes of our world operate according to the laws of energy.

DEFINITION OF A SYSTEM

The word *system* refers to anything that functions as a whole by the interaction of organized parts.

For example, a house is a system of water pipes, electrical wires, rooms, building materials, and so on. A radio is composed of various tubes, wires, and other electrical parts interconnected to operate as a whole. A football team consists of separate players operating in unison according to various invisible connections having to do with their training and motivation. A classroom is a system: it works only when there is a unified effort by teacher and students, with all members contributing and responding back and forth so that motivations develop and learning takes place. A forest is an ecological system consisting of trees, soils, chemical cycles, wildlife, and microorganisms interacting so that the forest as a whole is sustained. Galaxies are systems of stars that interact in their exchanges of matter, energy, and forces of gravity.

The human body is a system of organs that work together because the organization of blood vessels, nerves, digestive parts, muscles, and bones is maintained. Each of the major divisions of the body constitutes a subsystem. Thus, textbooks on the body discuss the digestive system, the circulatory system, the nervous system, etc. When we go into greater detail we find that these subsystems in turn have component cells and tissues which are also systems, since they too have parts, the microscopic components of living cells. There are systems within systems within systems. Since our minds cannot consider everything at once, we must decide on what *scale* to consider a subject or question. Most of the systems discussed in this text are the larger ones: environmental, economic, and social systems. The principles of energy and systems, however, also apply to small systems such as chemical reactions and to large systems like stars. Principles of energy and systems are a part of biology, physics, chemistry, industrial studies, and many other fields.

It is common, in our educational system, to begin consideration of a subject by examining parts, looking smaller and smaller. In this book, however, the reverse is true. We shall be putting parts together to see how whole systems work and how larger systems affect smaller ones. It is sometimes said that in order to understand a problem, we must look one size larger. That is, to understand one business, we have to look at the economy of a whole town. To understand a lake, we have to understand the region that exchanges river water and chemicals with it. To understand one nation, we have to consider the world.

SYSTEMS DIAGRAMS AND ENERGY FLOWS

Symbols

Figure 1-1 is a diagram of some processes of a farm. A farm is a system, since it has interacting parts. The square formed by the dashed lines shows the boundary of the system. Outside and within this boundary are shown some of the main interacting parts affecting the overall production of food by the farm. The plants are shown receiving materials and associated energy from outside sources that flow into the system along the pathways indicated by lines. Flowing out from the system is the food produced. Flowing downward is energy that has been used and is now in the form of dispensed heat: this is indicated by a special type of arrow. The diagram shows at a glance that

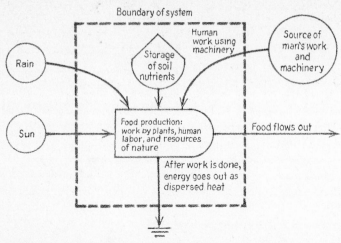

Boundary of system

Rain

Sun

Storage of soil nutrients

Human work using machinery

Source of man's work and machinery

Food production: work by plants, human labor, and resources of nature

Food flows out

After work is done, energy goes out as dispersed heat

Figure 1-1 Energy flows necessary for a farm to produce food. (See text and Figure 1-4 for explanation of symbols.)

production of food requires the steady interaction of inflows of sunlight, rain, nutrients stored in the soil, and inputs from human work using machinery.

Circles are used to designate the sources of inflows from outside the boundary. The symbol shaped like a water tank is used to indicate storage of soil nutrients within the boundary. The bullet-shaped symbol represents the interacting processes in the farm. Arrows show the direction of inflows and outflows. The arrow that points downward represents the outflow of used energy.

Each pathway represents a different kind of flow, but each is accompanied by a flow of energy. Each outside source and each storage tank is a place where energy is avilable to flow. Thus we call the systems diagram an *energy-flow diagram:* it shows the flow of energy along with flows of materials, money, information, and so forth.

Forces and Energy Pathways

Everyone has a concept of force as a causal action, as when one throws a ball, lifts a book, or sends a message. We use the word force in many ways, referring to economic forces, political forces, physical forces, etc. In physical science single forces, such as the force required to lift a book against the pull of gravity, are identified and measured. In our complex world, of course, there are many kinds of forces operating together in most phenomena. But whether we are talking about a single force involved in lifting a book or about the thousands of forces involved in the processes of a farm, these causal actions—single or in groups—require the flow of energy. Energy moves in the direction of the driving forces; forces and energy flows move together on the same pathways of causal action. For example, the pathway lines in the diagram of the farm

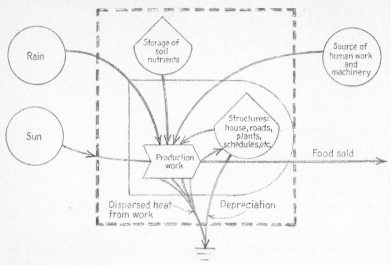

Figure 1-2 The farm shown in Figure 1-1, but with farm storages and interaction of outside energy sources included.

(Figure 1-1) are paths of energy flows and also lines of causal action. Four of the pathways (sun, rain, soil nutrients, and human work) come together; this indicates that forces of four types are interacting to produce food.

Interaction and Storage Symbols

Figure 1-2 is the same diagram as Figure 1-1, but shown with more detail about the interactions of the inflows to produce food. A pointed block is shown inside the farm symbol: it indicates that each of the inflows of food making is necessary to develop the work of the farm structure—planting crops and producing food. One arrow leads from the farm symbol into a storage symbol: here is a storage of the energy in structures of work—house, roads, plants, schedules, etc. These stored assets are also necessary to the main work; this is indicated by a pathway of work contribution that goes back from the storage tank to the left. In all, five pathways of interacting work and materials are required for the food-production process.

Note the downward pathway of loss of the stored structure by depreciation. There is always some depreciation of any storage, and losses due to depreciation must be shown on any diagram drawn to keep track of gains and losses in energy. The downward pathway in Figure 1-2 has two parts: the line on the left is the loss of heat energy in the production process; the line on the right is depreciation from stored energy.

Money

Figure 1-3 shows the same diagram again, but with addition of money. The flow of money is indicated by a dashed line. Money is shown coming into the

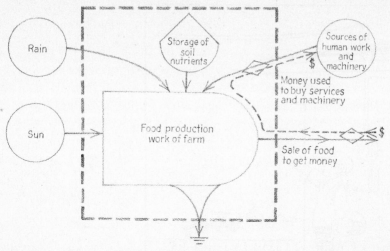

Figure 1-3   The farm shown in Figures 1-1 and 1-2, but here the money flow is included.

farm in exchange for sale of food and then going out (to the right) in exchange for fertilizer and machinery. The diamond-shaped symbol linking the money pathway to the food and machinery pathways indicates an exchange of money for goods.

Figure 1-4 includes all the symbols we have used so far. Although at first these symbols and pathways will be unfamiliar, you can see that they summarize our common experience in observing what a farm does and how it connects with weather (a system of nature) and with the economy (a human system).

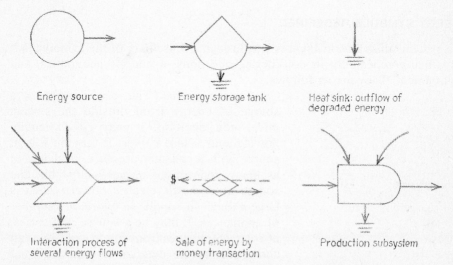

Figure 1-4   Summary of energy symbols introduced in this chapter.

"ELECTRICITY IS QUITE SIMPLE. IT FLOWS IN ON ONE LINE, AND THE MONEY FLOWS OUT ON THE OTHER."

DUNAGIN'S PEOPLE by Ralph Dunagin. Courtesy of Field News-paper Syndicate.

ENERGY SYMBOLS REDEFINED

The systems diagrams in Figures 1-1 through 1-4 used six of the symbols that we will use repeatedly in considering questions of energy, economics, and environment. They are as follows:

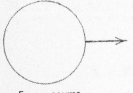

Energy source

Energy source. The circle indicates a source of energy from outside the system under consideration. It may be a steadily flowing source like a river. It may be a large source with a constant pressure, available to as many connections as necessary—like the source of domestic electric power, which is large enough to supply an enormous number of appliances. It may be a source that varies, as solar energy does from day to night. We can add words to the diagram to describe what kind of energy is being considered and how it is delivered.

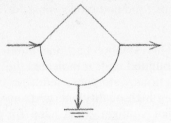

Energy storage tank

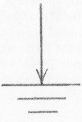

Heat sink

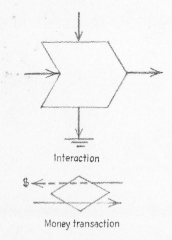

Interaction

Money transaction

Energy storage tank. This symbol indicates a storage of some kind of energy within the system. The symbol could indicate energy stored in an elevated water tank, in an oil tank, in the manufactured structures of a building, in a library (information), or in any way that makes it ordered and valuable.

Heat sink. The arrow pointing downward, seemingly into the ground, symbolizes the loss of degraded energy—that is, energy which cannot do any more work—from the system. The pathway of degraded energy flowing out includes heat energy that is degraded as a by-product of work and also the dispersal energy of depreciation.

Heat is lost in friction, as when an automobile's tires rotate over the road. It is also lost from automobile exhaust. In the human body, heat flows out from the skin and lungs.

Concentrations of matter are energy storages. Energy is lost if the concentration is spread apart: this is depreciation. Depreciation of an automobile is the loss of energy as it rusts, becomes worn, and gradually falls apart.

Heat sinks are required on all storage-tank symbols and all interaction symbols.

Interaction. The pointed block is used to show the interaction of two or more types of energy required for a process. In the example of the farm, sunlight interacts with water, soil, bought machinery, and stored structure: all these are required for the interaction that produces food.

Money transaction. The diamond-shaped symbol indicates the flow of money in one direction to pay for the flow of energy or energy-containing materials in the reverse direction. In Figure 1-3 we used the money-transaction symbol to indicate money obtained by the sale of food and money paid for machinery.

Production. This symbol has one blunt end and one the rounded end. It indicates the processes, interactions, storages, etc., involved in producing high-quality energy from a dilute source like sunlight. It is used for producer subsystems such as plants. The symbol can be used without interior details, as in Figure 1-1, or with interior details, as in Figure 1-2. In both figures, the symbol refers to the structures and processes of the farm, which is a subsystem producing food. This symbol is also used for natural ecological systems like forests, coral reefs, and biological communities of streams.

Production

PUTTING NUMBERS ON THE DIAGRAM

In Figure 1-5 some numbers have been put on the diagram of the farm. These represent the number of units of energy flowing in or flowing out each day. If the flows vary from time to time, we may show average values in order to view the quantities and how they interact. Each kind of flow could be expressed in its own unit of measure, such as pounds of fertilizer, inches of rain, footcandles of sunlight, dollars worth of machinery, and pounds of food. In Figure 1-5, however, all the flows are expressed in the same terms: Calories per day. Using a common unit of measure helps us to understand the comparative flows, and the Calorie is the unit that will be used to measure energy.

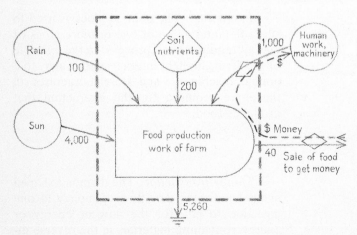

Figure 1-5 The farm with Calories of energy flow per day indicated on the pathways.

SUMMARY

Both public policy and individual attitudes reflect a basic confusion concerning the relationships of energy, economics, and environment and their influence on the drastic changes taking place in our lives in the 1970s. Fundamental questions are being raised about the future—about the continuation of growth and prosperity and the effects of inflation. To understand these issues (as we must do if we are to face the future with some confidence), we must understand the principles of energy, economics, and environment. Simple energy-flow diagrams can be of help. These allow us to see the behavior of whole systems in terms of the interactions of resources and processes.

We are now able to think about a farm as a system of flows and interacting parts as well as in terms of the familiar images of crops, tractors, farmhouses, and buying and selling. And we are beginning to get used to energy symbols as a way of expressing familiar ideas and putting them together in a way that is easy to visualize and remember. We should be able to think of the world around us as systems of various kinds, some contained within others.

Sometimes the paths that unite the parts are strong and visible, like wires connecting power lines and the pipes transporting natural gas. But sometimes the pathways are not visible until we put them down on paper: examples are actions of men, economic exchanges of buying and selling, and chemical cycles in our environment.

We can now begin to associate storages as important parts of systems. In spite of depreciation and losses, we see that these storages are kept renewed by the continuous inflow of energy from outside sources. We see also that several kinds of inputs interact to form structures and products to maintain the farm and to send out food from it.

Chapter 2

What Is Energy?

As we learned from examining the flows of interaction on a farm (Figure 1-2), everything that happens is an expression of the flow of energy in one of its forms. Other examples of flows of energy are the motion of an automobile, the storage of water in a tank elevated against gravity, the motions of water waves, the activities of ordinary housekeeping, the calculations of a computer, the education of human beings, and human religious feelings. All of these forms of energy flow can be related; for most forms, there are conversion factors which indicate how much of one kind of energy is equivalent to how much of another kind of energy. Most people use the word *energy* for inputs to their bodies or their economy, and thus think about energy as food, fuels, electric power, atomic power, and so on. However, components of energy are necessary for the action of all the processes of the universe. To understand the energy basis of man and nature, we must learn how energy is necessary to everything we do.

Energy is a measure of everything. It measures the amount of stored capability for future processes and the rate at which processes go. The total amount of an accomplished process is measured by the energy used.

Energy comes in from the sun as light and is received on earth, where it heats waters, produces plant food, and indirectly generates winds, waves, and the coal and petroleum in the ground. Everything has a component of energy.

The complicated interactions of resources in our factories, farms, and human activities are made up of combinations of flows of energy in various forms. Even the processing of information, as in books and by television, involves whatever energy is necessary to hold and transmit information. Often the energy responsible for some useful work is processed far away from the final work itself. The electric power we use for household appliances and lights is usually processed in huge, noisy plants many miles away. We are often unaware of how much energy forms the basis of our lives.

Diagrams can be used to summarize the main flows of energy, where they come in, and where they go out. Figure 1-1 showed that the energy for food production comes in with sunlight, rain, the work of the farmer, etc. Most of this incoming energy leaves the system as heat energy which spreads out into the surroundings. But the food that goes out of this system is high-grade energy, useful to man.

WHAT IS HEAT ENERGY?

Heat is familiar to everyone, for we all know how it feels to be hot or, in the absence of heat, cold. The concentration of heat can, of course, be measured with a thermometer, which contains a fluid that expands as the temperature rises. Heat is one of many forms that energy can take and is the one form into which all the other forms of energy can be fully converted. It is for this reason that we use a measure of heat, the Calorie, to express amounts of energy.

To understand what happens when matter is heated, we must understand that it is made up of units—atoms and molecules—which are connected in orderly patterns to make solids but can vibrate and make limited motions. In gases, atoms and molecules are free and move like balls bouncing off the walls of a billiard table or off each other. The atoms and molecules exchange motions with each other as they vibrate, rotate, and bounce around. The energy in the motion of these atoms and molecules is the energy that we feel and measure as heat. When the atoms and molecules of a substance are very active, the material is hot and the rapid bumping and vibrating motions of the atoms and molecules are transferred to the substances of your hand. This motion activates the nerves, so that you get a message from your hand that it is hot. If this motion is hot enough it damages the substances of your hand, producing a burn. In summary, heat is the energy of the mixed, chaotic motion of atoms and molecules in all directions (Figure 2-1).

Energy in the form of heat is hard to harness because the molecular motions are chaotic. Heat is degraded energy if everything is at the same temperature. If there are differences in temperature, the chaotic motions tend to spread from points of concentration (high temperature) to points of lesser concentration (lower temperature). Thus if there are *differences* in temperature, heat energy can cause processes, such as the flow of heat.

The world we live in requires a certain level of heat for the best operation of its components. People, animals, ecological components, cities, etc., work best at a range of temperature that may require the generation of heat.

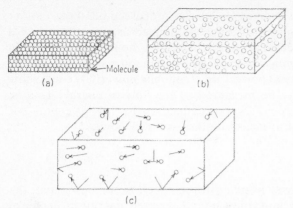

Figure 2-1 Atoms and molecules whose motions constitute heat, one of the forms of energy: (*a*) solid, (*b*) liquid, (*c*) gas.

CALORIES—MEASURE OF ALL ENERGY

Since all kinds of energy can be converted into heat, we can measure the energy flowing into and out of a system in units of heat energy such as the Calorie.[1] Most of us are familiar with the Calorie in connection with nutrition and know that the human being requires about 2,500 Calories a day.

The Calorie is the unit used in Figure 1-5 to express the inflows and outflows of energy. The numbers shown on the pathways indicate the amount of energy, in Calories, derived from the various inflows. The total inflow is balanced by the total outflow. All processes require energy in some form— light, motion, magnetism, electricity, chemistry, etc. *Because all forms of energy can be converted into heat, energy can be defined and measured as the ability to generate heat.* The unit of measure, as has been stated, is the Calorie.[2]

To fully characterize energy for the purposes of determining its usefulness we must consider how concentrated it is; later, we will consider measures of the ability of energy to do work. Although the Calorie will be used as the main unit of energy in this text, many other units are also a part of common experience. One reason why it may be difficult to comprehend that energy is in everything is that different units of energy are used in different areas of our technology. A similar situation exists when we consider units of length: we have meters, feet, yards, inches, miles, etc., and we must somehow learn to think in terms of one or two of these units and convert to the others as necessary.

Calorie Equivalents of Other Units of Energy

We can change all kinds of units of energy into Calories: the Btu, the foot-pound, the joule, the watt-hour, etc.[3]

Power

Power is the rate at which energy flows. The power flow in a pathway is the Calories passing over it per day. For example, the power of the flow of food energy in a human being is the number of Calories processed per day. Another familiar unit of power is horsepower, originally defined as rate of work of a normal-size horse. Rate of electrical energy used by a household is given in terms of kilowatts. The word *power* is also used in a general way to indicate the ability to influence, as in "political power" or "economic power." We will show that even these processes may be measured by the flow of energy. That is, physical and biological measures of power, defined as the flow of energy per unit of time, may also be used to measure other kinds of power.[4]

KINDS OF ENERGY

Energy, as we have said, comes in many forms, such as sunlight, heat, the stored energy of an elevated tank of water, the stored energy in chemical substances, electrical energy, magnetic energy, the momentum of a moving vehicle, and so on. Energy of one kind may be transferred into energy of other kinds if there is a suitable conversion process. There are natural laws governing the amount of energy that can be converted to more highly concentrated energy, as we will see in Chapter 3. Let us now consider various forms of energy.

The sun drives the flows of energy on earth. Directly or indirectly, most of the flows of energy on earth come from the sun. Directly or indirectly, then, sunlight maintains man and nature on the earth. Taking place within the sun are atomic reactions something like those of a hydrogen bomb; on the sun, however, the great gravitational force holds the explosion in. The energy of the atomic reactions ultimately goes out from the sun as light, which is in the form of little packets of energy called *photons.* As the photons recede from the sun, they spread out so that by the time they reach the earth, they are relatively far apart. The sunlight received by the earth is, therefore, diluted: relatively few photons arriving at any one place at any one time. (See Figure 2-2.) When sunlight reaches land or water, the land or water is heated because the light is transformed into another form of energy, heat. Sunlight also reaches green plants, which use it in complex processes with water and soil nutrients to make food (see Figure 1-1). The food matter produced by plants is used by natural ecological systems for maintenance and growth, and also by farms to support human beings.

In certain places, such as swamps, the mouths of rivers, and briny seas, some organic matter made by plants is covered over by sand and sediment before it can be eaten by animals or microorganisms. Under the conditions of heat and pressure that accompany the state of being covered over for thousands or millions of years, such organic matter is converted to coal and oil. These deposits show signs of the original plant matter in a fossilized state and are consequently called *fossil fuel.* The process of conversion, working slowly over time, created large storages of coal and oil. In the last century, these storages have been pumped and mined to support the human culture of cities, machines, power plants, automobiles, and the like. But we are apparently using

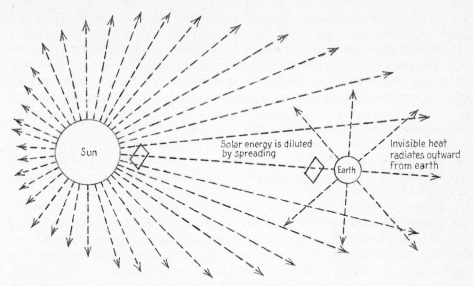

Figure 2-2 Energy from the sun heats the earth, and energy leaves the earth as heat radiation.

the fossil fuels faster than they are being generated by the processes accompanying burial of organic matter. There will eventually come a time when fossil fuels that are near the ground surface are exhausted. Indeed, we already have to go so deep in the earth and so far out to sea to find fossil fuels that they are no longer a cheap way of supporting human activities.

Energy from the sun that is not converted into food heats the land and seas. *Temperature* is the measure of concentration of heat. Thus where there is a difference in temperature, there is a difference in heat concentration. Concentrated heat energy tends to wander from hot areas toward cold areas—that is, toward areas where heat energy is less concentrated. When two areas have evened out the original difference in heat distribution, so that both are at the same temperature, the energy that existed in the form of uneven concentrations of heat has been *degraded*.

When the sun heats parts of the earth, differences in temperature are generated, as between land and sea and between the tropics and Arctic. Differences in temperature cause winds, as when hot air rises. When water is heated by the sun, some of it evaporates into the air. This water vapor in the air is transported by the wind over land, where it may fall as rain or snow. The energy of water caught behind mountain dams was originally received from the rain. Such elevated water is a form of stored energy that can be used to turn waterwheels and turbines on its way back to the sea. Water makes a cycle from oceans to rain to land to rivers and back to the oceans. We often harness water power and wind power. The sun-driven water cycle also wears down mountains, depositing sediment at the coast. The weight of sediment helps to make buried organic matter into fossil fuel, as has been noted.

Ocean waves and ocean currents are another form of energy. These are partly generated from the winds and thus are ultimately driven by the sun.

In most places in the world's oceans, the tide rises and falls once or twice a day. The tides are created by the pull of the moon on the sea as the earth turns. Whatever started the moon and earth spinning around the sun gave them an initial energy that is gradually being transferred into tidal motions. Tidal energy is used by ecosystems and geological processes to maintain the patterns at the sea bottoms and landscapes along the shores.

Electrical energy is the energy in concentrations of electrons, which are charged atomic particles. Electrical energy is generated by special processes like thunderstorms and power generators. Nuclear energy is the energy stored in the nuclei of atoms.

Figure 2-1 showed how matter is composed of atoms and molecules that are rigidly organized in solids, flow over each other while remaining in contact in liquids, and bound about loosely in gasses. In Figure 2-1 the atoms are shown as balls. But in fact atoms have an interior structure: an atom consists of a central part, called the atomic *nucleus,* which contains many kinds of atomic particles, and a surrounding part made of charged particles called *electrons.* The external electrons of an atom are held near the nucleus much as the planets are held near the sun by its gravity.

The binding of atomic particles contains stored energy that can sometimes be transformed into other forms of energy. Energy in the outside electrons is involved when atoms combine to form compounds and interact in chemical reactions (see Figure 2-3). Much greater energy is stored in the atomic nucleus; it is this energy that is released as sudden surges of heat when an atomic bomb is exploded.

When a nuclear reaction takes place, as in the sun or in an atomic bomb, there is a release from storage of nuclear energy as heat, light, and speeding nuclear particles. Nuclear reactions take place under controlled conditions in some nuclear power plants so that very concentrated heat is released. The very hot core thus produced provides a large temperature difference which is used to operate steam engines and turbines.

Any object that is moving relative to a place has kinetic energy relative to that place. A moving automobile has such kinetic energy, and the amount of energy rises as the square of the speed. If rapidly moving automobiles collide, the energy stored in the motion is converted immediately into work of destruction and into heat. Before the collision the materials of the automobile were all moving together; we call such motion *kinetic energy.* After the crash, the motion has been converted to the random vibrations and motions of the metal molecules. The tangled metal resulting is hot to the touch—no energy was lost; it was converted into heat.

Processes Requiring Interactions of Energies

When two kinds of energy are both necessary for a process, the final output depends on both. For example, two materials may react in a chemical process, as when oxygen reacts to burn wood. The inflowing energy is both in the substances and in their relationship. In the example just given, since both oxygen and wood are necessary, we must regard each as carrying some part of the energy that is transformed in the process. We may diagram this by showing two pathways meeting in a common block (see Figures 1-4 and 2-3). We may also consider either flow as making the other go. Each flow, by turning the other one on, becomes a multiplier. Many people think that a flow of material such as the oxygen for a fire is not energy but only raw material. It is both, however: any material essential to an energy process has a component of energy.

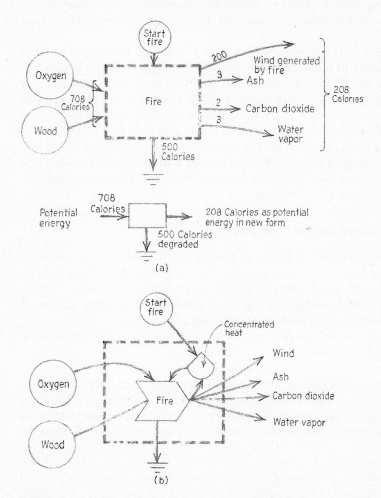

Figure 2-3 Energy diagram of the reaction of a forest fire:

  Oxygen   +   wood   <u>fire</u>→ wind + carbon dioxide + water vapor +   ash
192 grams   190 grams   264 grams   108 grams   10 grams
(a) Calorie equivalents of inflowing and outflowing energy. (b) Energy relationships necessary for the energy flow in a fire.

The interaction of oxygen and wood in a fire is shown in Figure 2-3; it produces fire winds, with carbon dioxide, water vapor, and dispersed heat as the other outflows. This example illustrates the nature of interactive energy flows. In Figure 2-3a the inflowing substances (wood and oxygen) each carry some chemical energy. These inflowing energies are shown to be equal to the total outflows of energy including heat, wind energy, and the chemical energy still stored in the waste products (carbon dioxide and water vapor).[5] Because oxygen in the air is abundant, in this example wood is the component in short supply. Still, neither component would be able to release its energy without the other's reacting with it. The fact that one component is necessary to release the energy of the other is indicated by the interaction symbol (see Figure 2-3b).

In the example of the farm in Figure 1-1 we indicated the participation of three kinds of energy in the overall conversion of sunlight to food. Figure 1-2, which gives more detail, indicates the *necessity* of all three kinds of energy by putting the energy pathways on a pointed-block symbol rather than running them together as though they were interchangeable and could be added. Figure 1-2 also shows the dispersion of much energy into degraded heat. The work of the structure that is maintained by the output process is another necessity. Thus, Figure 1-2 recognizes many types of work and energy as necessary to produce food: rain, sun, soil nutrients, human work and machinery, and the feedback from the farm's own storage.

QUALITY AND CONCENTRATION OF ENERGY

As has already been stated, all forms of energy can be converted completely into heat, which is low-quality energy. Heat-energy equivalents, Calories, can serve to measure all kinds of energy. In any system, inflowing Calorie equivalents must equal outflowing Calorie equivalents, including dispersed heat. Calories measure energy when it is downgraded to heat.

A second measure of energy has to do with *upgrading.* It takes more energy of one kind to generate energy of another, higher-quality kind (Figure 2-4).

The various kinds of energy that we have just discussed differ in quality. Some forms of energy, like sunlight, are very dilute; others, like gasoline, dynamite, and high-voltage electricity, are very concentrated. A Calorie of dilute energy cannot be used in the same way as a Calorie of concentrated

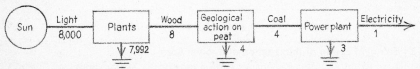

Figure 2-4 A scale of energy quality showing energy that must be degraded to upgrade each type of energy to the next level of quality.

energy. Furthermore, it takes energy to concentrate energy. We must degrade some energy in order to concentrate what is left. This can be seen in Figure 1-2, where sunlight, a dilute form of energy, interacts with some other sources of energy to produce food, a concentrated form of energy. In order to do this, part of the inflowing energy is degraded and dispersed in an unusable form. In considering sources of energy, many people fail to recognize that the concentrated energies needed for the intense activities of cities themselves require much concentrating of energy. Many Calories of dilute energy are needed to form one Calorie of concentrated energy.

There is a scale of quality of energy in which forms of energy that are least concentrated appear at one end and forms of energy that are most concentrated appear at the other end. To convert a dilute form to a more concentrated form, there must be a degradation of a considerable part of the energy. Four Calories of coal are required for a Calorie of household electricity; 1,000 Calories of sunlight may be required to make one Calorie of wood. Figure 2-4 gives a scale of quality of energy and some of the conversion factors for going from one form of energy to another. These factors include the energy cost of any machinery that the conversion process may require.

Energies which differ in quality differ in their ability to do work. When work is done, and the energy used in doing it flows into dispersed heat, we may estimate the amount of work done by the Calories of heat energy that went through the system. However, the ability to do work cannot be determined from the Calories of energy available unless we also know the quality of the energy. A Calorie of dispersed heat cannot do any work. A Calorie of dilute sunlight must be concentrated to do much work. But a Calorie of fuel or atomic energy has a very high concentration of energy. Concentrated Calories can do more work, drive more processes, and involve more forces as they go from concentrated form to the form of dispersed heat. (See Figure 2-4 for different qualities of energy and the amount of other types of energy necessary to generate certain qualities.) However, to do more work they must interact as amplifiers to flows of lower-quality energy.

Efficiency

Any ratio of energy flows is called an *efficiency.* The most important efficiency is the ratio of desired energy output to all of energy input. For example, the ratio of wood production to sunlight inflow in Figure 2-4 is 8:8000 or 0.1 percent. Different kinds of energy have different efficiencies. The efficiency of conversion of Calories of a low quality to Calories of a higher quality is a measure of the usefulness of the higher-quality type of energy. For example, 1 Calorie of electricity is generated for every 3.4 Calories of coal burned in a power plant and burned elsewhere in other industries to keep the power plant supplied with equipment and services.

Many kinds of high-quality energy are required for complex work. We tend to think of the energy requirements of a process only as *fuel,* ignoring human work and the contributions of materials. Actually, however, the energy expended in obtaining the material to be used as fuels—mining, manufacturing, processing, storing, and transportation—may be larger than the fuels themselves.

Consider the energy required to run an automobile. Much more is required than the fuel (gasoline): there are the energies involved in manufacturing the machinery, in supplying replacement parts, in keeping the driver and the repairman trained to operate the automobile, and in supplying the roads to run it on. Figure 2-5 is an energy diagram that shows the flow of the requirements for an automobile. Notice that the fuel is only one part of the energy ultimately responsible for the automobile. The energy flows of materials and work must be included in the energy cost of a system and its operation.

People sometimes make mistakes in deciding what activities ultimately cost or do not cost much energy, because they do not consider all the energies actually involved. Activities such as education seem like small energy consumers because they seem to involve only people and not many fuel-using machines. In fact, however, the energy involved in the long chain of converging educational activities is very large.

We are so used to thinking of energy as physical processes that we do not realize that thinking uses energy too. The energy is in all the work that goes into educating the mind and maintaining the body to support the mind. The reading you are doing and the thinking you are doing about what you are reading are a use of energy. Because much energy is used in developing one's abilities,

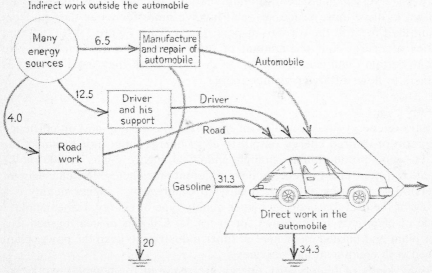

Figure 2-5 Many hidden energy flows are required to operate an automobile that is driven 10,000 miles per year, and some of these are located far away from it. Numbers, in millions of Calories per year, are based on data assembled by E. Hirst (1973).

intellectual activity is a very high-quality use of energy; intelligence and learning are concentrated potential energy. Flows of energy do work and also cause all kinds of activities not ordinarily thought of as work, such as the enjoyment of beauty and the feeling of love.

SUMMARY

We have mentioned many kinds of energy and discussed the transformation of energy from one type to another. Most of the flows of energy on the earth are directly or indirectly derived from the sun's rays. Photons of light separate from each other as they leave the sun, becoming ever weaker in concentration as they move toward the earth.

Interactions of two or more forms of energy in a process involve multiplier actions between flows necessary to the process. Where a chain of energy transformations is long and varied, it is easy to lose sight of the ultimate cost in energy of a process, especially if the final flow that the energy supports is delicate and of high quality.

Energy was defined as a quantity that flows through all processes, measured by its ability to generate heat. The rate of flow of energy was defined as *power.*

The *concentration* of energy affects the amount of useful work that can be done. Concentrated energy does more work by helping less concentrated energy flows. Most of the energy becomes less concentrated during the work process. Most of its ability to do work is used up.

By now, the reader should sense that everything has a component of energy and that flows of energy form the basis for nature and human life. Chapter 3 restates the aspects we have just discussed as the laws and principles of energy.

FOOTNOTES

1   The word *Calorie* spelled with a capital C means a *kilocalorie,* sometimes called a *large Calorie.* It is equal to 1,000 small calories. A small calorie is the heat energy required to raise the temperature of 1 milliliter of water 1 degree Centigrade when the water is at 15° Centigrade.
2   Many physics courses start out with the definition of energy as the ability to do work. This is true if one is considering forms of energy of the same quality. But Calories of energy of different qualities do different amounts of work. A Calorie of dispersed heat cannot do any work. It seems better, then, to avoid the definition of energy as work until we explain quality of energy.
3   One large Calorie of heat is generated by the work of 3.97 British thermal units (btu), 3,087 foot-pounds, 4,186 joules, and 1.17 watt-hours. One kilowatt-hour generates 860 large Calories. To make mental comparisons, learn the approximate conversion factors: 1 large Calorie is released by or is equivalent in energy units to 4 btu, 3,000 foot-pounds, 4,000 joules, and 1 watt-hour. A barrel of oil releases about 1,600,000 kilocalories when it does its work.

4   An energy flow of 1 Calorie per day is 5.61 horsepower, or 4.186 kilowatts.

5   The chemical potential energy is estimated for 1 mole of carbohydrate and assigned equally to the inflowing oxygen and wood. The kinetic energy of wind and the dispersion of degraded energy as heat are outflowing; some of the heat is absorbed as water evaporates. Five percent of the wood may be mineral elements that remain as ash. There are small amounts of potential energy in the remaining concentrations of carbon dioxide, water vapor, and ash. The evaporation of 108 grams of water absorbs 65 Calories of heat, but this heat only changes state from liquid to vapor. The fire is cooler because of the water evaporated, but this is not part of the main budget of potential energy which can be used to do other work.

Chapter 3

Principles of Energy Flows

There are several natural laws of energy that apparently have no exception on earth. These laws give us an understanding of the limits to human life and nature on our planet. We must learn these laws of energy in order to develop common sense about our future, about plans and propositions we are offered, and about how energy affects our money. Energy diagrams such as Figure 1-3 may be helpful in making the laws of energy clear. We have already discussed the behavior of energy as it flows according to these laws and principles; this chapter, however, will state the principles more concretely. These laws and principles are widely believed to be among the most important laws of the universe. They are presented here more as they would be in a science textbook.

PRINCIPLE 1: LAW OF CONSERVATION OF ENERGY

The first law of energy is the *law of conservation of energy:* The energy entering a system must be accounted for either as being stored there or as flowing out. Energy is neither created nor destroyed.[1] For example, in Figure 1-5 energy flows in from several sources, including that flowing with materials, and is accounted for as outflowing food and degraded (waste) heat energy. Ultimately, the energy that comes to earth from the sun leaves the earth in the form of invisible heat radiation (see Figure 2-1). All objects send out rays of energy called *radiation.* Rays from very hot objects are visible to the eye; we call them *light.* Rays from warm objects are not visible; they are sometimes called *heat radiation.* You have probably felt heat rays when walking near a warm wall. The energy flowing out from the earth as heat radiation equals the energy flowing in from sunlight, except for some temporary seasonal storages.

PRINCIPLE 2: LAW OF DEGRADATION OF ENERGY

The second law is the *law of degradation of energy:* In all processes some of the energy loses its ability to do work and is degraded in quality. In Figures 1-5 and 3-1 the necessary degradation and dispersal of used energy as waste heat are indicated by an arrow into the ground. We call this the *heat-sink symbol.* We keep the second law of energy in mind by putting the heat-sink symbol on the diagram for every process.

Energy that has the ability to do work is called *potential energy* and is useful; energy that has done work is degraded and is no longer useful. When most people refer to the energy supply—gasoline—for the work of an automobile, they really are referring to potential energy. When people say energy has been "used up," they mean that potential energy capable of doing work has been converted into a degraded form of energy, usually dispersed heat.

Thus the second law of energy is a familiar idea to most people. We are used to the idea that foods and fuels cannot be used more than once. Since energy that is potentially capable of doing work cannot do so again after the work is done, potential energy cannot be used over and over. Although inflowing energy equals outflowing energy, by the first law of energy, most energy loses its ability to do work after any work processes. For example, in Figure 1-5 the work of food production results in most of the potential energy's going into the process and being degraded into heat energy given off at a low temperature into the environment. Some energy is retained after the work as high-quality food that goes out for consumption. According to the second law of energy, the amount of high-grade energy resulting from the production process is less than the energy coming in. Any process must degrade some of its energy.

Another way to think of the second principle of energy is to recognize that dispersed heat is random motion of molecules: the motions of the molecules are so minute that they cannot be seen, but they can be felt as the sensation that we call *heat.* The processes of our world start with potential energy in concentrated form and go toward dispersed heat energy. The tendency of organized "stuff" to wander into randomness is what makes work processes, and ultimately all energy processes, go.

Another way to describe the tendency for potential energy to be used up and degraded is to say that *entropy* always increases in real processes. Entropy is a technical measure of the amount of disorder[2], but the word is also frequently used in everyday conversation.

Order and Disorder

Everyone has a sense of order and disorder. Books neatly arranged on a shelf are orderly; books scattered through a room are disorderly. Threads woven into a rug are orderly; a jumble of unwoven threads is disorderly. A building is orderly; the raw materials in the work lot before construction are relatively unordered. Words in a book are ordered; words jumbled, as in a word game, are disorderly. When human beings set up social relationships, order is developed in mental programs and memories.

Another way of stating the second principle of energy is to say that orderly structure, patterns, and arrangements tend to drift toward disorderly. The concentration and arrangement tend to come apart. Any concentrated and organized situation has a potential energy relative to a disorderly state, and this energy tends to depreciate as order is lost and dispersed. Since everything that is ordered depreciates, maintaining order requires supplying more potential energy to do special ordering work—to reweave a rug, rearrange books, rewrite words, reprint books, reestablish social relationships.

The flow of energy through the biosphere is continually fostering the building of order by arranging patterns from disordered raw materials. The disordered materials are put together into new orders. Such disordered materials are obtained from recycling waste products of things that have depreciated beyond effective use. Recycling of materials is a regular part of natural systems such as forests and seas. It is also a part of human systems, especially when they have no outside (imported) sources of rich energy concentrations and reserves.

Figure 3-1 (page 40) is an order-disorder diagram showing the overall buildup and breakdown of a system. For example, on the farm, buildings, crops, etc., are produced from disordered materials. Then they go back into disorder, the buildings through deterioration and the crops by being eaten and digested. Energy is used to build the order. Energy is degraded and exported as the order breaks down. The order built by energy flows represents an increase in the quality of energy.

PRINCIPLE 3: SYSTEMS WHICH USE ENERGY BEST SURVIVE

The third principle of energy, called the *maximum-power principle,* explains why certain systems survive: That system survives which gets most energy and uses energy most effectively in competition with other systems.

For example, consider a farm like that in Figure 1-1. A farm which plants crops at the best time in relation to rain and sun, which uses the best fertilizer to make the crops grow, and which grows crops which people will buy, is the one which will produce enough so that the farmer lives well and can repeat the process year after year. The successful farmer's system will survive and will then be copied by other farmers. Farms with patterns that maximize flows of energy survive.

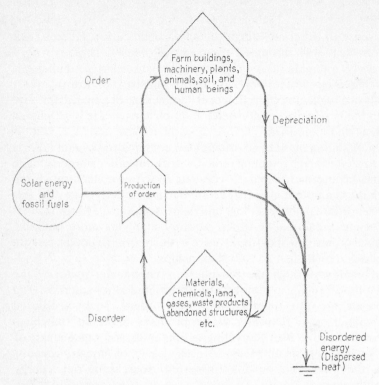

Figure 3-1 Order-disorder cycle on a farm.

This principle of competition for the best use of energy explains why certain energy systems survive others. We are used to this principle as it applies to business competition. Consider tire manufacturers: the company that builds the best tire for the least money will survive because it sells more and uses its income to buy additional energy inflows. A successful business maximizes its flow of energy and does more.

The maximum-power principle also controls forests. The kinds and numbers of trees which will take over in a particular forest are those which best use the energy of the sun, the rain, and the soil to maintain more forest activity. Other types and combinations die out.

Another example of maximizing energy for survival is the early American colonies. The colonists who used the energies available in the new land, plus energy-based goods they had brought with them, plus energy-based learning of how to put resources together for the survival of the group, were those who lived and spread.

The maximum-power principle may also be stated as follows: *Those systems that survive in the competition among alternative choices are those that develop more power inflow and use it best to meet the needs of survival.* They do this by: (1) developing storages of high-quality energy, (2) feeding back work from the storages to increase inflows; (3) recycling materials as needed; (4) organizing control mechanisms that keep the system adapted and stable; and (5) setting up exchanges with other systems to supply special energy needs.

Feedback Loop of High-Quality Energy

We have just said that systems which survive are those which maximize their uses of inflowing energy resources. Maximizing the use of inflowing energy involves storing up some high-quality structures, information, technology, abilities, etc. These storages must be capable of increasing the flow of energy and keeping the system stable in the long run. We may diagram these ideas as in Figure 3-2. An energy source, *E*, is available to a system where some of it is upgraded in the process at *A* into a flow of quality products, *B*, that are organized as storage and structure, *C*. *C* is shown interacting back through *D* on the energy inflows at *A* to pump in more energy and continue the process. When there is enough flow of *E*, growth of *C* continues up to the point where inflows balance outflows. A steady state comes when the maintenance costs balance the inflow energies from production.

Notice that the pathway of energy flow that has the storage of quality structure, *C*, doing work in interaction with the main energy inflows, *E*, is an arrow, *D*, going to the left in the opposite direction from the main energy flow, *B*, which runs to the right. When a flow goes back from a point downstream in the energy path to interact at an upstream point, it is called a *feedback*. The principle that systems win out in competition by maximizing their energy resources implies at least one feedback pathway of work in each system. As Figure 3-2 shows, a closed loop is formed between the energy flow, *B*, and the feedback, *D*, that returns actions back upstream.

Feedbacks are found throughout all systems of man and nature. Feedback

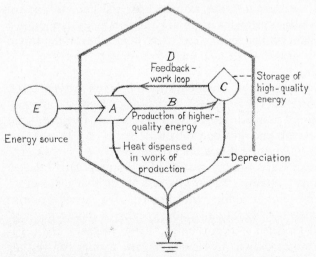

Figure 3-2 Defining main pathways of self-maintaining units with feedback loops.

loops must form if a system is to survive in the continuous competition among systems to maintain their energy supplies in the best way possible. There is always variation among systems, so that there is always choice; the best system survives and is copied by others.

The feedback loop is symbiotic. Flow *B* makes possible the return work, *D*. The feedback loop *D* in turn rewards the upstream components *A* with quality work. The closed loop (*B* and *D*) is a symbiosis between upstream and downstream: that is, upstream and downstream are necessary to each other. Whatever energy is degraded in the process of developing the downstream storage is made up for by the special high-quality action of the return feedback. If the effect were not fully mutual, the system would not do as well at obtaining energy or would waste energy.

CHARACTERISTIC STORAGE SYSTEMS

Figure 3-3*a* illustrates the characteristic pattern that tends to develop because of the principles of energy. By the first principle (conservation of energy) the 100 Calories of energy flowing in per day flow out per day—90 dispersed in the main production process and 10 in the depreciation of the storage. By the second principle of energy (degradation of energy) each process along the energy flow disperses some energy into heat that is no longer usable for work. The first process is production of order, and the second is storing of high-quality energy. By the third principle (maximization of energy), power is maximized by developing a storage of high-quality energy with a feedback of work to make possible the maximum inflow of energy. Some of the storage energy is used to pump in more energy from outside sources.

The world is full of subsystems with one or more of the characteristic features: storage, depreciation pathways, and feedback interactions. Figure 3-3*b* and *c* illustrate two systems that are examples of the general pattern shown in Figure 3-3*a*: a town that provides machinery and services to operate a dam and a hydroelectric plant that generates electrical power to run the town; and a sheep that lives on grass. Both systems store energy and use it to facilitate getting more energy.

Another example of a system with a feedback loop is given in Figure 1-2; the farm has storages of structure (plant tissues, farmhouses, people, information, and farm animals) which work to catch sunlight, water, fertilizer, and other inputs in order to develop enough food to sell.

Systems which survive build order to aid the use of energy. We have stated that feedback loops are necessary for the survival of systems. High-quality storages are also important. Systems that survive are those that build more patterns of useful structure. For example, the pattern of a surviving farm includes buildings, schedules for operation, roads, means of marketing, and economic know-how—and the farmers know this. All the patterns and organized materials such as those shown in Figure 1-2 are means for further operation. Some such order must be developed in any process, if the system is

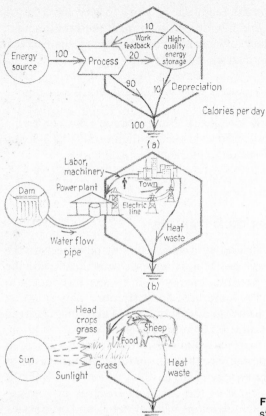

Figure 3-3 Characteristic patterns of storage, feedback, and maintenance that maximize survival energy.

to compete successfully in getting and using energy. According to the second principle of energy, energy must be degraded; but according to this third principle, storages of concentrated energy must be built and maintained if a system is to succeed. For example, in Figure 3-3*b*, order is in the buildings and social structure of the town; in Figure 3-3*c*, order is in the body of the sheep.

In the biosphere, weather systems develop. For example, thunderstorm winds bring in more energy of the winds' moist air to make the storm powerful. The storm then becomes the dominant weather in the area. Order is in the pattern of clouds and vertical distribution of air masses.

SYMBOL FOR SELF-MAINTAINING ACTIVITY

The three principles of energy require that any unit in our world of systems and subsystems must balance inflowing energy and outflowing energy in order to survive. It must degrade some energy and also build some high-quality ordered structures and feed its work back to aid the energy flow. The principles thus require at least some storage and some feedback of work to interact with the incoming energy, as shown in Figure 3-2. Units which meet the requirements are self-maintaining: they use their own resources to operate themselves. Examples are organisms, towns, industries, and humanity.

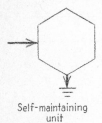

Self-maintaining
unit

We need a general abbreviation—a symbol—for units which implies that there are storages and feedback interactions: the hexagon, shown here, is used in Figures 3-2 and 3-3, and in subsequent diagrams.

FLOWS OF ENERGY TURN CYCLES OF MATERIALS

Figure 3-4*a* shows a water wheel driven by a steady inflow of water. As the water drops from a higher to a lower position, it releases some heat while the wheel goes round and round. The potential energy of the water has gone into the energy of motion as well as into some heat from friction. This situation illustrates a common property of our world of systems, large and small. Energy flows in one direction from a concentrated condition to a dispersed condition, and while it does its work, it turns materials round and round. Energy is stored in the wheel as kinetic energy while the wheel turns, and the turning is part of the means of collecting more water energy more easily.

Figure 3-4*b* shows the water cycle of the earth. The energy of sunlight heats the seas, especially in the tropics, develops wind systems, and evaporates water which circulates over the land and the polar regions. The water drops back to the land in the form of rain and snow. Then, by glaciers and rivers, it returns to the sea. Energy flows in continuously and flows out as degraded, less-usable heat energy; in this process, it turns the water cycle around. There is energy stored temporarily in the water as it is raised into the air, and this energy helps return the water to the sea where it is available to go around again. There is also potential chemical energy in the purity of freshwater relative to saltwater. The water cycle is like the wheel that continuously gets an impetus of energy to continue its steady turning. The cycle of water also helps capture more of the sun's energy and thus helps the system of nature to survive.

Another example is a balanced aquarium in the schoolroom. Such an aquarium is in many ways a model of the biosphere. Light energy flows in to provide light and heat, and then leaves the system as dispersed heat. The plants bind the raw materials to make food. Parts of the plants and the microorganisms consume the stored food, returning the raw materials to the environment. In this way the materials make a complete circle, while the energy passes through the system and out. Again, there is a closed circle of materials turning like a wheel under the impetus of the steady flow of energy.

Figure 3-4*c* shows the biosphere, which does the same thing. The

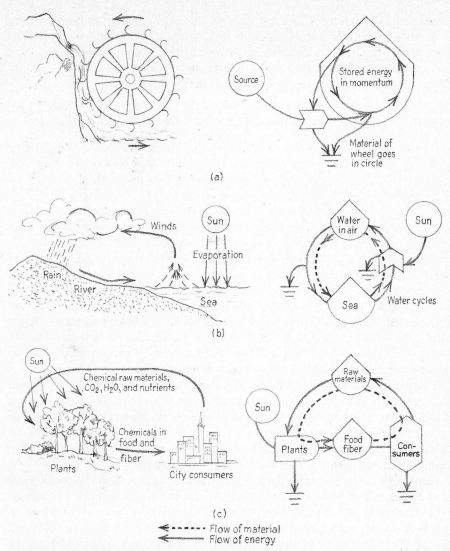

(a)

(b)

(c)

<<< - - - - - Flow of material
<<<————— Flow of energy

Figure 3-4 Flows of energy drive the cycle of materials. (*a*) Water wheels. (*b*) Water-rain–runoff cycle. (*c*) Chemical cycle between plants and consumers.

biosphere takes sunlight to form the food that natural forests and marine organisms provide to their consumers as well as the food from agricultural production that goes to human beings. Then these foods and fibers are utilized by consumers, including human beings, animals, cities, microorganisms, and fire. The consumption returns the raw materials to the environment for reuse. The materials in this case are those involved in the growth of plants and in human nutrition. They are the carbon dioxide of the air; the water of the rain; and the fertilizer chemicals, especially nitrogen, phosphorus, and potash. The cycling of all these elements is driven by the flowing in and out of the energy of sunlight as it keeps the plants, animals, and cities of the world healthy. The turning of the material cycles is a regular part of the processing of energy. The cycles store energies as they go, and their patterns make energy flow steadily and the life of the biosphere continues harmoniously.

SOURCES OF ENERGY CONTROL THE DESIGN OF SYSTEMS

A system is ultimately controlled by sources of energy outside it. An estuary with a tide develops entirely different landscapes and organisms from one without a tide. A system rich in solar energy has a different pattern of vegetation from one poor in solar energy because of cloudiness or a high latitude. The patterns of agriculture that developed in ancient civilizations were based on sun and rain alone. In the industrialized agriculture of many countries today, there are the additional sources of fossil fuel that directly and indirectly supply machinery and services for farming. Consequently, ancient patterns differ from modern ones.

Competition for survival leads each system to be different from others if its combination of available energy sources is different. In the same area, on newly cleared land, one farmer plants corn, another wheat, and so on. After several years, if other factors remain the same, all the farmers will probably be planting the same crops: they will have discovered which combination of planting produces crops that sell for the most money.

External sources of energy are the basis for a system; and the system gradually fits itself, its storages, its material cycles, its feedbacks, and its design to that pattern which maximizes energy in the combination available to it. In the process of trial and error, there is a selection from among the choices that develop as a result of random variations. Surviving systems are those that feed back their stored energy to stimulate the flow of energy.

When a situation changes, it takes time for a system to develop that fits the new energy sources well. For example, when climates change, new kinds of vegetation invade and replace the previous ones. When there are changes in patterns of energy, other changes follow in agriculture, industry, culture, and styles of life. The nature of the sources of energy sometimes helps us to see what the new patterns may be. For example, when fossil fuel became available, transportation became more important, and "automobile cities" developed.

Limited and Unlimited Sources of Energy

Whether a system grows or not depends on whether its source of energy can support further growth or is limited to the extent that it can maintain only a small amount of activity. A large dam is relatively unlimited in supplying equal water pressure to as many small water pipes as might be connected. A large pool of oil is relatively unlimited to the first few oil wells connected. A large power plant can supply relatively unlimited electricity to all the first electric appliances plugged in. One may add users to supplies like these until the source does become limiting (Figure 3-5*a*).

Steady-flowing but limited supplies of energy (Figure 3-5*b*) stand in contrast to the virtually unlimited supplies. Examples are the flow of a small creek turning a water wheel, oil flowing from a seepage crack, and electricity from a small generator. In these examples, the user cannot get any more energy than is regularly made available. Pumping will *not* generate any more energy than is flowing regularly per unit of time. The flow is limited at the source. The word *demand* is sometimes used for consumers' attempts to get more energy. With limited energy sources, efforts to meet an increased demand for energy eventually fail. No users can be added beyond those using the energy as fast as it is supplied. Sunlight is another example. A forest cannot get any more energy than comes in regularly per acre per day. Once it has built enough leaves to catch all the available light, it can do no more to maximize the energy flowing in from this source.

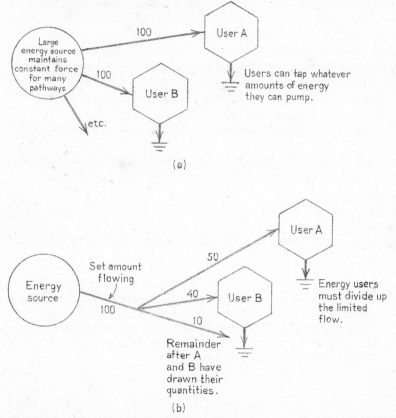

Figure 3-5 Comparison of force-controlled and flow-controlled energy sources. (*a*) Large energy sources with a force on each user. (*b*) Limited energy source with only a set amount flowing and available per unit of time.

The difference between the two kinds of sources may be indicated in our systems diagrams, as shown in Figure 3-5. Figure 3-5a shows the unlimited source; this is indicated by a line drawn directly from the circle. Such a source can provide forces along whatever pathway may be connected, as with the first appliances plugged into a large power plant. In Figure 3-5b the source gives an energy flow which is limited; this is indicated by the pathway with a fixed flow. Users of limited energy must hook into the limited flow. An example is the earthworms of a forest floor: they depend on the steady dropping of leaves and are limited to a rate of dropping which is controlled at the source—in the limbs of the trees.

SUMMARY

There are three main principles of energetics: (1) the law of conservation of energy, (2) the law of eventual degradation of energy, and (3) the maximum-power principle. To survive under competition, systems build order, develop feedbacks of energy, and recycle materials. These principles and the systems of energy flows that result from them, can be diagramed as a characteristic, basic, recurring pattern. High-quality energy from system storages pumps energy from sources of energy through a production unit where more high-quality energy is generated. The production process sends energy into the storages to keep up with losses due to depreciation, losses due to the work of pumping, and losses due to export to other systems, if any.

Unlimited energy sources can support increasing consumption of energy and accumulations of more storages that we call *growth*. Source-limited energy flows cannot support unlimited growth, and the systems using such sources stop growing at the level of storage that the limited inflow can support.

FOOTNOTES

1 Energy can be generated from mass in atomic processes, like atomic explosions, but these processes are rare on earth.
2 Entropy is defined by the amount of heat involved in changing from a condition of maximum order of no molecular motion at zero temperature (absolute zero) on the kelvin scale to an observed condition containing heat. Entropy change is the ratio of change in Calories of heat to the temperature in kelvins. The kelvin temperature scale is degrees centigrade plus 273°. The ratio of heat to temperature drops where heat flows out and increases where heat is added, but the total rise in entropy is greater than its loss when both are considered.

Chapter 4

Energy and Money

Everyone is familiar with money and with buying and selling; most people use money to judge the values of products for sale in stores. Everyone is becoming interested in energy, too, especially since it has begun to be in short supply as growth begins to catch up with available resources. Money flows in circles, but energy flows through a system and ultimately out in a degraded form no longer capable of driving work. The flow of energy makes possible the circulation of money, and the manipulation of money can control the flow of energy. We must understand something about money and energy and their relationship in order to understand the economic system and the way energy affects it. In this chapter money is included in the diagrams and the relationship of money to energy is shown. We will discuss how the availability of external energy controls the circulation of money and what happens in inflation. Both energy and money will be used to measure value.

THE MONEY CYCLE

Energy and money flow in opposite directions. As food produced on a farm goes to a town, the townspeople pay the farmer money that goes back to the farm. The farmer uses money to buy machinery and fertilizer from the town,

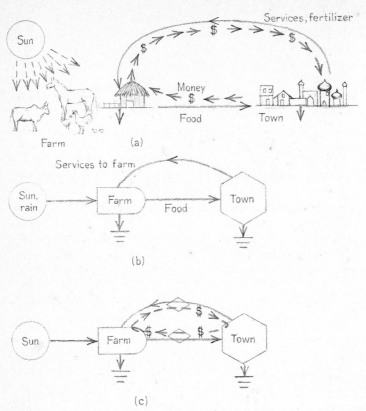

Figure 4-1 Relation of money cycle to energy flow in the exchanges between farms and a town. (*a*) Exchange between a primitive farm and a town, as in India. (*b*) Flow of energy. (*c*) Money flowing as a countercurrent to energy loop.

sending the money back to town to pay for it. As Figure 4-1 shows, these relationships form a circle: money goes around and around; energy flows in as high-grade potential energy and is used to maintain the structures of farm and town, but much of it necessarily goes out as low-grade dispersed heat.

Transactions involving money require work. Much energy goes for business activity, paperwork, banking, etc. The money-transaction symbol first used in Chapter 1 indicates that some of the potential energy flowing is being used for such work.

Human economic systems can bring in materials and fuels to support populations and cultures. However, human beings are only a small part of the great biosphere of oceans, atmosphere, mountains, valleys, land, rivers, forests, and ecological components. Ultimately it is not just human beings and their money that determine what is important; it is all the world's energy. It is, therefore, a mistake to measure everything by money. Instead, we should use energy as the measure, since only in this way can we account for the contribution of nature. For example, money received by a farmer for his crops pays only for his human work—not for the work of rain, soil, wind, and sun. We take natural forces for granted until one is gone. Figure 1-3 shows the various kinds of energy that go into a farmer's crops, including those that he must buy, such as work and machinery, and those that come free from the biosphere.

In fact, the flow of energy turns the cycle of money. Money passing through a human economy is an example of a cycle driven by and dependent on the steady inflow of energy. The money cycle, however, turns in opposite direction from the usual cycles of matter. Money passing from hand to hand is like ball bearings that turn in the opposite direction from the main wheel: by their turning, they make the wheel turn better, with less friction. Figure 4-1a shows the cycle of money for a primitive agricultural economy, such as once existed in India, based mainly on the sun's energy. Food from farms goes to town, and the services of people and human wastes ("night soil") come back to the farms, completing the material cycles and facilitating the work of the farms.

When money is added as a medium of economic exchange, it goes in the opposite direction. The townspeople pay the farmers for food and the farmers

"The heck with the money! Let's lock up the food!"

Copyright © 1973 by The Chicago Sun-Times. Reproduced by courtesy of Wil-Jo Associates, Inc., and Bill Mauldin.

in turn pay the townspeople for the fertilizer and services they give back to him. Money flows in a circle turning in the opposite direction from the flow of energy; because it helps account for services given and value received, it facilitates the system of material flows and the receipt and processing of energy. Energy stored in a financial system is in the form of information, money, know-how, and social agreements. These have to be maintained at some cost of potential energy, but they make the overall system of cycles run better and capture more energy so that the system can compete successfully with alternative systems. For example, a money system is more effective than a barter system: it is driven by the flow of energy and builds special structures and cycles so as to be the best possible processor of energy.

To understand the relationship of money and energy one must recognize that money is a cycle which does not turn unless material cycles are turning and energy is flowing. If energy becomes less, the system must turn more slowly. Like all the other flows, money helps work processes go; but the money flow itself causes work and must thus be regarded as having its component of energy too.

Ratio of Energy to Money

Money tends to be spent when people have it, and it goes around money-circulation loops like the one shown in Figure 4-1. Money can go around only if energy flows through the system to support the work that the money buys. The more work is done for each dollar that circulates, the more truly valuable the dollar is. At any given time there is an average ratio of money flow to energy. For example, Figure 4-2 shows $1.4 trillion ($10^{12}$) circulating per year in the United States. During 1973 about 35 million billion Calories of energy were used in the United States and degraded into dispersed heat. The ratio of these two flows is 25,000 Calories: $1. Naturally, this ratio varies in different parts of the economy, but we may estimate the ratio for the system as a whole.

This ratio is very helpful in showing how much energy is being used to support economic activity. Suppose a person makes and spends $10,000 a year. Multiplying this by 25,000 Calories (remember that the ratio of energy to money is 25,000 Calories: $1) we find that 250,000,000 Calories of work is done yearly in support of that person. Now since his personal budget for food energy for the year is only about 1,000,000 Calories, the difference represents work by farm machines, power plants, industry, and nature. Such work done in support of the human being is very large; but since much of it is far removed from the individual, he is often not aware of it.

"Externals": Sources of Energy

As is shown Figure 4-2, flows of money go in complete circles. When money goes out of one industry as purchases it must come back in with sales—if the business is to continue. A *balance of payments* is required for each part of the economic system. Many ideas about economics have to do with stimulating or retarding the circulation of money in order to stimulate the production of real

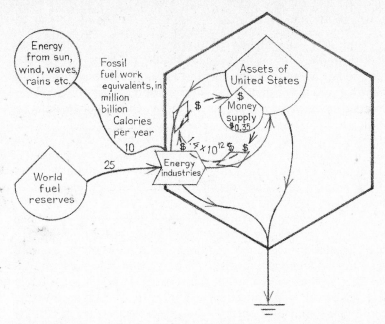

Figure 4-2 Energy flow and money circulation in the United States in 1973.

value. But the real basis of the economic system is *outside* the money circle. Economists refer to these inflows from outside as "externals": they are, in fact, sources of energy. As long as the externals of a system are constant in effect, the circulation of money is a principal factor controlling the economy. (In Figure 4-2, the money circle controls the flow of energy so long as the external energy sources are constant in effect.) However, when the effect of externals is not constant—when the availability of energy changes—the economy is changed in a way that overrides and is mainly independent of manipulations of money within the circle. Figure 4-2 shows that as the external source of energy labeled "world fuel reserves" diminishes, the relationship of money to energy within the system changes.

Turnover

In any money economy a quantity of money is circulating. At any place in the money circle, the rate of spending tends to increase when money on hand increases. Money tends to circulate rapidly: in our system, it goes around about four times a year: that is, the *turnover rate* is four times a year. For example, if there is $250 billion in the economy, $250 billion is spent in a quarter of a year and $1 trillion is spent in a year. The turnover of money is illustrated by Figure 4-2, where the quantity of money represented in the dollar storage tank is $350 billion and the rate of spending is $1.4 trillion per year. While money is circulating, energy is being pumped in the opposite direction.

INFLATION

The *buying power* of money is the amount of real goods and services that it can buy. If the value of the money diminishes, so that a dollar buys less, we call this *inflation*. Inflation can be caused by increasing the amount of money circulating without increasing the amount of energy flowing and doing work. In Germany after World War I so much money was printed that each bill became almost worthless; it took a wheelbarrow of money to buy a loaf of bread. Inflation can also be caused by decreasing the amount of work being done without decreasing the amount of money circulating. When energy is scarce, so that it is not easy to augment human work by the work of machines, the amount of work decreases while the amount of money circulating remains the same. Consequently, a dollar buys less work and is therefore worth less.

In recent years, the federal government of the United States has been deliberately increasing the amount of money circulating in order to make sure that the economic circulation was good. Adding a little more money caused more to be spent, allowed some new projects to be started, and thus spurred some growth. As long as there was unused fuel energy to be tapped, adding money did stimulate growth and caused new energy to be drawn into the economy to do the work. Because energy was available as needed, adding money increased the demand for energy and increased the energy flowing, so that the economy grew. Adding money also caused some inflation—3 to 5 percent a year. The effect was to decrease the value of people's savings while stimulating the economy to grow. Inflation was like a tax: people's money was converted into new government projects.

In Figure 4-3, inflation is graphed over a three-century period. As is shown, there is a tendency for prices to rise during wars. In wartime, much energy is diverted from normal production into military action and destruction, which do not immediately generate value. Since energy is diverted while the amount of money circulating is the same or greater owing to government's financing the war, the energy per dollar in the main economy decreases. By 1940 there was a general inflationary trend caused by the practice of increasing money to stimulate flows of energy, and by the gradually increasing work required to process each unit of energy as the richest sources of energy were used up.

DEPRESSION AND RECESSION

During the depression which began in 1929, the business and financial world of money, banks, loans, stocks, etc., was so upset that the circulation of money slowed down. People held what money they had in reserve to make it last longer. With little money flowing, there was little energy flowing. The production of goods and services slowed down, and growth stopped. A *depression* is defined as the slowing down of the circulation of money so that productivity and inflows of energy are slowed down. In the life of our nation there have been alternating periods of depression and good times. In 1930 the economy was so

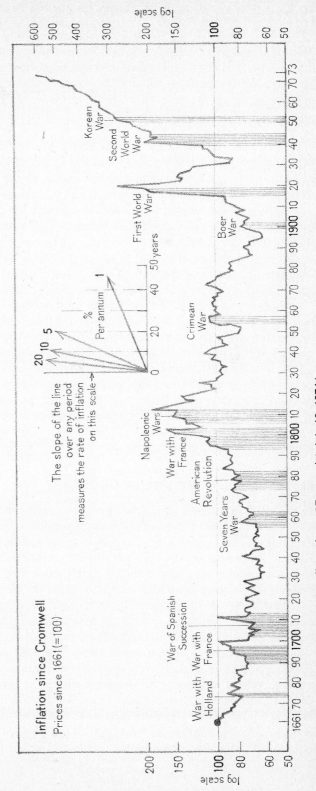

Figure 4-3 Record of inflation and the effect of war. (*Economist*, July 13, 1974.)

disrupted that there were few jobs, and for those without jobs life was terrible. The trouble was not a shortage of energy, but a shortage of circulating money, a shortage of institutions to process money, and a psychological inhibition against spending. Massive efforts were made by the government to put money into circulation, stimulate the circulation of money, and revive production processes and flows of energy. These efforts were only partly successful. This was also a period when farmers moved to the city, each with a period of low productivity during transition. It was the outbreak of World War II, with the need to work hard for survival and a change from the pattern of holding money to one of spending it, that finally ended the depression.

Around 1974, there was inflation created by decreasing energy. Since money was losing value rapidly (inflation in 1974 was 12 percent), people spent it as fast as they could in order to convert it into valuable goods and services before it lost value. Thus bad times in the 1970s have been characterized by diminishing inflows of energy and a stopping of growth; but people did not stop spending at the approach of hard times. Money did not stop flowing.

Stopping of growth and decline in total productivity, work, and economic activity are sometimes called a *recession*. (In this book, the words *recession* and *depression* are used interchangeably.) The recession of the 1970s has been entirely different in cause and effect from the recession of the 1930s. In the 1930s, a badly functioning economic system and a transition from a rural solar economy to an urban fossil-fuel economy created lower productivity than was possible and desirable. This was, of course, a bad time for many people. Whether a recession due to declining energy reserves must produce such bad times or whether it can lead to a steadier economy and more favorable times is a question of great importance.

EFFECT OF A LIMITED INFLOW OF ENERGY

In the 1930s, stimulating the flow of money could stimulate the inflow of energy and work because there was abundant available energy. In the 1970s, rich deposits of energy-carrying materials such as petroleum, copper, and fertilizer have become scarce. The problem, then, is a shortage of available energy, and therefore money may not stimulate much more flow of energy. When the rate of inflowing energy is as high as it can be, more money cannot increase it.

In Chapter 3 we showed the difference between an unlimited source of energy (like a dam large enough to drive any number of water wheels) and a flow-limited source (like a small creek that can turn only one water wheel). When the energy available to our country was a large reservoir of oil available to be tapped as needed, then increasing the circulation of money pumped more energy. When, however, the available energy is flowing in at a set rate that cannot be increased, then measures aimed at circulating money faster can bring no new energy. The only effect is to make the money less valuable in relation to energy. Money represents less real work, and we have only stimulated inflation. Figure 4-4 compares the two situations. Figure 4-4*a* shows a large,

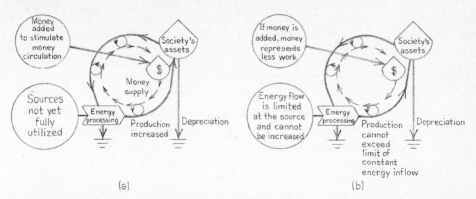

Figure 4-4 Comparison of the ability of money to stimulate the economy in the 1930s and the 1970s. (*a*) 1930s: unlimited energy, financial machinery stalled. (*b*) 1970s: pumping from a limited source.

relatively unlimited source of energy. The more money is increased in this situation, the faster the spending wheel turns, and the more energy flows out of the source. In Figure 4-4*b*, however, the flow of energy cannot be increased, because its rate of supply is limited at the source. Attempts to pump more by increasing money and the rate of spending money must fail. Energy flows at the same rate. The money flow, however, is increased, and the value of the money is therefore reduced, since each dollar buys less energy. People with savings are hurt because they can buy less with the money they have saved than they could have bought at the time they earned it.

CAPITAL ASSETS

In the diagrams that we have used so far, the systems shown have internal storages of structure that include buildings, people, food stocks, information, culture, education, memories, and all things that we regard as useful, valuable, and subject to depreciation if unprotected. All these stored items are part of our assets; we sometimes call them our *capital* assets. (See Figure 4-2.) It is from such storages that we draw the means to continue old activities, pump in more energy, and start new activities. Capital assets accumulate when the inflow of productive work exceeds usage, depreciations, and other outflows. (See the flows in Figure 4-2.) Capital assets accumulate during growth and are maintained only by means of a continual inflow of additions to replace the outflows from storage. Accumulating the products of growth is done by, for example, an estuary in growing fish, by a forest in developing a stand of trees, and by a human economy in developing buildings, more people, and the various urban assets.

If a system has the energy to grow and does grow, developing the storages that we call *capital assets*, it now has a storage of energy and can put more feedback work into pumping in new energy. An economy with storages of energy can add money while keeping the ratio of money to energy constant.

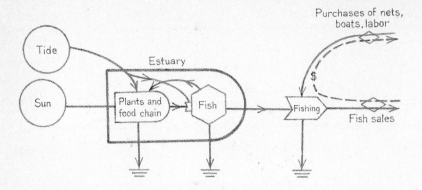

Figure 4-5 Work of an estuary supplies fish; money is not involved until fish are caught.

Money accumulated along with storages of real energy is called *capital money,* and it can be loaned. Extra storages of energy can be enlisted in new activities. The storages of energy are processed by advancing capital money as loans to the people doing the new activities; that is, this money is used to buy the stored energies for the new activities.

ENERGY FLOWS WITHOUT MONEY FLOWS

Money is exchanged between people as a medium for keeping track of the exchange of goods and services. However, money is not exchanged between the parts of a natural ecosystem (like a forest). For example, Figure 4-5 shows the system that supports fishermen supplying fish for commerce. Most of the energy involved in developing this valuable resource is in food chains based on solar energy, in work of tides, and in weather systems. Money is involved only in the last stages of the fisherman's harvest and sale. Therefore, money in this case measures only the work of the fisherman, not the work of the estuary. Money is not a measure of value here. Energy, however, is a measure of all the contributions to the product, including those of the fisherman, those involving paid services, and those of nature.

SUMMARY

Money has now been added to our consideration of the systems of flows of energy that support our existence. Money circulates in the opposite direction from energy, facilitating the flow of energy. The ratio of energy flowing to money flowing is useful, since it measures inflation and the value of money in purchasing goods and services.

Since the energy involved in work is an unchanging measure of what has been accomplished, energy is found to be the best measure of value. The energy value represented by money, however, varies with the ratio of energy to money circulating. Money is inadequate as a measure of value, since much of

DOONESBURY

Universal Press Syndicate.

the valuable work upon which the biosphere depends is done by ecological systems, atmospheric systems, and geological systems that do not involve money.

Inhibiting the circulation of money slows down the flow of energy, as in the depression of the 1930s. Stimulation of the circulation of money by adding money will stimulate the flow of energy only when supplies of energy are large. Adding money when sources of energy are limited merely creates inflation.

Chapter 5

Energy and Growth

As we face the future, all of us are interested in how energy may cause farms, cities, and whole economies to grow, level off, or vary in other ways. Everyone has observed that some things grow and others waste away. Some things remain fairly constant; others vary regularly; still others vary in a random way although they may be characterized by an average state or a constant trend. It is variation in sources and storages of energy that may produce changing or unchanging patterns. Will our economy grow, or has it reached a steady state? What about the world? The reader will be able to make better predictions after finishing this chapter, which discusses the forces on which growth depends. We will consider some of the main ways that flows of energy may control growth, stability, and other patterns that change with time.

Again, systems diagrams will be used to describe flows, storages, and interactions of energy. When we simplify a complicated situation into a diagram as we have been doing, the simple diagram is called a *model*.[1] We will learn in this chapter which graph shapes go with our models. Six basic models are discussed, starting with the simple tank and ending with a model that may best describe our present world situation. Two examples are given for each model: a natural ecological system and a human economic system.

GRAPHS OF QUANTITIES OVER TIME

Because we are discussing variations with time, each of the energy situations presented here will be accompanied by a graph, which shows variations over time of quantities of order stored in the system. Most readers are familiar with graphs of time, such as graphs showing growth of population or variations in prices. But for readers who may be unfamiliar with graphs, a brief review follows. Figure 5-1 is an example; it shows the growth of population in the United States over time. Time is represented by the left-to-right (horizontal) axis; population, by the bottom-to-top (vertical) axis. The higher the graph line goes, the greater the population. Units of quantity are marked along the vertical axis. If you trace the graph line from left to right, you are tracing the changes in population as time passes.

Another example is Figure 5-2, a graph of stored materials such as leaves on a forest floor or water stored in a tank. As the graph line goes from left to right it records what happens from one hour to the next. The time is marked along the lower horizontal axis of the graph. The first time shown is just after the start of filling the tank; the quantity of water rises at first and then levels off so that with the further passage of time there is no further change.

DEFINITIONS: GROWTH, DECLINE, AND STEADY STATE

Depending on the availability of energy and the condition of the system at the start, a storage may grow, reach a state of unchanging size, or decline. *Growth* is defined as a period of expanding storages of structures, reserves of energy, population, information, or other aspects of order. In growth, inflows exceed outflows. *Decline* is a situation in which outflows exceed inflows; storages are decreasing. The state in which inflows just keep up with depreciation and losses is called a *steady state;* in this state, storages are constant. Flows into a storage may come from outside sources (as in Figure 5-3*a*) or from energy interaction processes (as in Figure 5-3*b*).

Examples of growth are the increase of weeds in a bare field left fallow; the increase in population, urban structures, and knowledge in the United States in the past century; and the rapid increase of microscopic algae in a lake which has been made rich by sewage waste nutrients.

An example of decline is a decaying log being eaten by a population of animals: as the log is eaten and gets smaller and smaller, the population of animals declines.

An example of a steady state is the population of plants and fishes in the constant-temperature springs of Florida. They grow and replace their losses without much change year after year. Other examples are a lawn in the tropics which is cut as fast as it grows so that the pattern remains the same; and the farms that used to exist in the Ganges valley of India, where the numbers of people, cows, farms, and institutions were fairly constant, growth replacing the death and repair replacing deterioration.

May 31, 1976.

Dear Kids,

We had such a good visit with you guys. Hospitality & food etc. were superb.

When I unpacked my suitcase I found Amy's red bootie. Sorry about that. I know they are the favorites. Outside of feeling depressed 'cause of missing you-know-who, we are feeling good. Lots of traffic & one accident (not ours) on the way home. All the campers were travelling early because of the rain. Well, at least we weren't on the road with the drunk and/or tired drivers.

Mary Jo & Tom had worked in their yard. It looks nice. They had us over for dinner. I guess they knew we'd be feeling low.

I lost 1/2 pound over the week-end.

Jim, will you do me a favor? Please get me that yellow print shell. It was 4.00

You can pay up to 6.00 if its not on sale anymore. The blouse was a "small". So I guess I can wear a small. But get a medium if there are no "smalls" No need to mail it.

Sorry to give you all the jobs. If you can't hack it, don't worry.

Jonathan learned to peddle his bike today.

Thanks again for a beautiful week-end.

June 15   Clematis blooming   Love   Mom + Pop

P.S. I know we forgot our album.

at cottage
June 21, 1976

sci teacher M.S.

Lenox

June 17 apple)

chem
anat
phys   micro bio

ELIZ   KENDALL
associate dean
Holliston Jr. Coll.
Lenox
Mass.

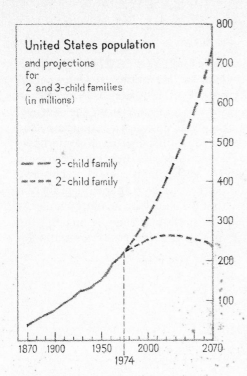

Figure 5-1 This example of a graph of growth over time shows the history of the United States population before 1974 with two predictions for the future. (*Time*, Sept. 16, 1974. Reprinted by permission from TIME, The Weekly Newsmagazine: Copyright Time Inc.)

ENERGY CONTROLS GROWTH

Growth curves depend on sources of energy and the use of energy. Sources of energy outside a system and storages of energy inside a system are responsible for the pressures that cause flows along pathways of use. (See Figure 3-5 and the accompanying discussion of limited and unlimited sources.) Sources of energy and the way storages of energy are used determine whether energy produces growth, a steady state, or a decline. The rest of this chapter will show some commonly observed situations in which energy determines growth curves. We will consider unlimited growth, source-limited growth, two kinds of self-pumping growth, an unrenewable source, and a combination of renewable and unrenewable sources.

To understand how the graphs come about, we can look at an energy diagram as though it were a system of water tanks, pipes, and pumps. Growth is very easily visualized from our experience in watching sinks and bathtubs fill or drain; and this common experience is an easy way to begin visualizing growth as it will be shown in the diagram. Moreover, even if you have difficulty visualizing, you can memorize the graph shape.

The models are another way of writing mathematics; one may write mathematical equations that describe the models and generate growth curves, and the equations for the models in this chapter are given on page 271. These models have also been simulated using computers to generate the graphs.

MODEL 1: GROWTH AND STEADY STATE IN A SINGLE STORAGE TANK WITHOUT FEEDBACK

The first model we show (Figure 5-2) is a simple tank. Its stored quantity grows or declines depending on the balance of inflows and outflows. One energy-storage tank is shown; there is a steady source of inflowing energy and an outflow that leaves the system according to the pressure from the storage. Some energy is always lost as dispersed heat (heat sink), as is required by the law of degradation of energy. Thus some energy is dispersed and some flows out with the materials. Do you see how a one-tank situation causes growth followed by a leveling off of growth and a subsequent steady state? The sequence of growth and leveling off would be the same whether for a grain bin, a migrant-labor camp, or an army. Suppose that we start with an empty tank. At first the inflow is filling the tank faster than the outflow is running out. But the outflow increases as the pressure of water in the tank increases, so that soon

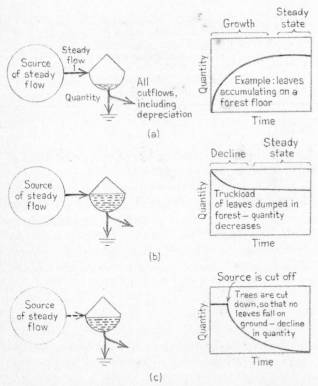

Figure 5-2 Model 1: Growth, steady state, and decline for a system of one storage tank and an energy source which maintains a constant flow. Examples: Leaves on a forest floor; a sink with water flowing in slowly and the drain open; immigrants coming to an island in a steady flow and leaving more and more as crowding increases. (*a*) Start with storage tank empty. (*b*) Start with full tank. (*c*) Start with steady state; then cut off energy source.

the inflow is balanced by the outflow. At this point, growth is replaced by a steady state (Figure 5-2*a*). Try this yourself by filling a sink with stopper out.

Suppose the tank had contained more water at the outset; then, there would be more outflow than could be maintained by a limited inflow and the tank would decrease from its initial state to a lower steady-state level (Figure 5-2*b*).

Suppose we turn off the source of energy. Now there is only an outflow; the tank begins to drain rapidly at first but slows down as the amount in the tank decreases, thus decreasing the pressure to flow out on the loss pathways. The graph of decline is a characteristic curved shape (Figure 5-2*c*).

Another example is leaves falling to the ground in a forest. Year after year they fall, and the amount of leaves on the ground builds up until the rate of decomposition equals the rate at which leaves drop. If the trees are removed, the source of leaves is cut off, and the storage of leaves on the ground decays away.

Storage and Outflow Pressure

The forces that push flows depend on storages. A tank of water pushes water through pipes according to the amount of water that piles up within the tank. This also applies to the heat sink: the more storage, the more depreciation. The flow is greater when a tank is full. In Figure 5-2*a*, as the tank fills, its storage exerts more pressures along the outflow pathways.

The dependence of flow on pressure holds true in other situations. For example, pressure for some political action depends on the number of people who agree on it; the rate at which a fire burns depends on how much firewood is in one place and ready to burn. Flow is proportional to force; this holds true for all simple pathways in the diagrams.

Source with Constant Flow and Source with Constant Force

The energy source in Figure 5-2 delivers a constant flow: the users are limited to this rate. Figure 5-3, on the other hand, shows a source that maintains a constant pressure, or constant force. For model 1, a difference in type of source is not very significant—a constant force applied to an unchanging inflow pathway delivers a constant flow. But for model 2, the type of source does make a difference—more energy can be pumped in.

Backforces

In a simple tank with a steady flow (Figure 5-2*a*), the tank has no effect on the inflow if the inflow is arranged so that there are no back pressures. Water added to a sink from a faucet above, for example, receives no back pressure from the water already in the sink. In our language of diagrams we use the arrowhead to indicate that inflows are arranged so that there are no back pressures.[2]

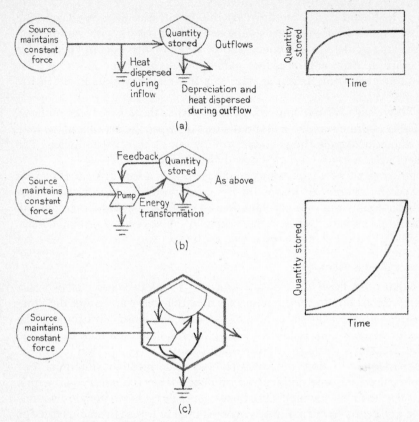

Figure 5-3 Effect of feedback pump on the acceleration of growth when the energy source maintains constant pressure. (*a*) Tank without feedback pump. (*b*) Model 2: Tank with feedback pump. (*c*) Tank with feedback pump, but with symbols clustered within hexagon. Examples of *b* and *c*: cancer, growth of colonizing population in a fertile new area, microbes in food.

MODEL 2: FEEDBACK-ACCELERATED GROWTH WITH LARGE SOURCE OF ENERGY

The single-tank system shown in Figure 5-3*a* produces growth that slows down and levels; other kinds of energy-flow situations produce entirely different growth curves. One of these is the accelerated growth that results when energy is fed back from the storage to accelerate the pumping in of energy from a source large enough to maintain its force. The storage shown in Figure 5-3*b* has a pathway of work that goes back to the left to pump in more energy. If, as in Figures 5-3 and 5-4*a*, the energy source is large, it can maintain a constant force. This is shown in the diagram by a direct line from the source circle to the interaction symbol: pumping the source can supply as much energy as is pumped. The more energy is pumped into the storage, the more is fed back. This energy-pumping feedback stimulates inflow from the source of energy.

The result is that the tank fills faster and faster without limit, at least until the ability of the source to maintain constant force is exceeded.

Suppose that stored water behind a large dam is generating electric power to run a village (Figure 3-3b). As the village grows, it feeds back its work to construct larger hydroelectric generators in order to tap more of the water pressure of the dam to drive more electricity so that the village can continue to grow. With more growth, more energy is fed back to add more power generators. Consequently, growth accelerates faster and faster, at least as long as the stored water in the dam maintains enough pressure to drive all the added electric generators.

Another example of accelerated growth is cancer. (See Figure 5-3b.) Cells grow; then each cell divides into two, which in turn divide, and so on, always increasing the quantity of cells. These cells feed back to the growth process by using more of the source of energy (the blood supply) to grow and divide again. This process will continue until something vital is interrupted. In situations of runaway pollution and food spoilage, many kinds of microbial cells grow this way.

The steeply rising graph of growth produced by the feedback acceleration with a large source is sometimes called *Malthusian growth*. Thomas Malthus (1766–1834), from whose name the term is derived, lived during the time when rapid growth of human population began because increasing energy was available as the industrial revolution started to use fossil fuel. Malthus feared that the rapid growth of population would exhaust resources and ultimately cause starvation and disease. Urban economies started off with the accelerated growth pattern of Figure 5-3b. Because reserves of fossil fuel were large, the predicted shortages did not develop until the 1970s. Now, since sources of energy are not maintaining constant force, feedback-accelerated growth can be only temporary. A better model for the past and future of humanity will be given at the end of the chapter (Figure 5-8).

Notice the shape of the growth curve in Figure 5-3b: it curves sharply upward. Contrast this with the unaccelerated growth curve in Figure 5-3a, which is steepest at first and then levels off. When there is feedback and the source of energy can maintain a constant force, as in Figure 5-3b and c, growth continues and does not level off into a steady state.

In the tank with the feedback pumping action (Figure 5-3b) the more the tank fills, the more it sends back effort at pumping in more. As long as the source is large enough to maintain a constant force in spite of a greater drain of energy, pumping will increase faster and faster. Since in reality no source is literally unlimited, model 2 is valid only for short periods. Limits do eventually develop in any actual situation.

A major difference in the nature of growth graphs results from differences in the arrangements for acquiring energy. In discussing the third principle of energy (the maximization of energy) in Chapter 3, we noted that the feedback arrangement is in general the one that leads to survival. A rapid upward curve like that in Figure 5-3*b* is characteristic of successful competition up to the time when sources of energy begin to reveal limits. The past growth of the United States may be an example, as we will consider further in Part Two.

MODEL 3: SOURCE-LIMITED GROWTH

Consider now what happens when the source of energy is not a large, unlimited one maintaining constant force, but rather supplies a constant flow of energy. An example is a stream of water supplying energy to a water wheel that generates electricity. In such a constant-inflow situation, one cannot use energy any faster than it is supplied to the source stream. There is only so much energy coming through per unit of time, and once this rate of energy is being used to the fullest, no more can be pumped. The flow is limited at the source. If one bases growth of a village on the harnessing of energy from such a stream, there can be only a brief period of growth of storage before the quantity of assets developed levels into a steady state (see Figure 5-4*b*). Compare this with an unlimited source (Figure 5-4*a*).

The limited-flow source is shown (Figure 5-4*b*) as a pathway out of which user systems must pump. If the user pumps less than is flowing in from the

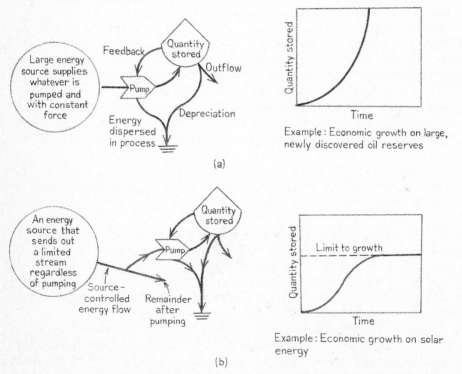

Example: Economic growth on large, newly discovered oil reserves

(a)

Example: Economic growth on solar energy

(b)

Figure 5-4 Effect of type of energy source on growth graph. (*a*) Force-maintained source. (*b*) Model 3: Limited-flow source.

stream, some amount flows on by as a remainder. If the user arranges to pump into use all the energy, there is no remainder.

Another example of source-limited growth is a forest growing on a regular inflow of sunlight. When the quantity of forest foliage has reached a level at which all the light energy is being used, growth levels off, as in Figure 5-4b.

Before the industrial revolution, human culture developed to levels that used all the regular available energy.Since the forms of energy were renewable and steady—from sun, wind, and water—growth was limited by the source and steady states often developed.

MODEL 4: SUPER-ACCELERATED GROWTH

The growth of assets (population, cities, and money) in the United States early in this century was actually faster than the Malthusian curve. Whereas the Malthusian curve results from pumping new production in proportion to the quantity of assets already stored, growth in the United States involved pumping at a faster rate than this. It was a *super-acceleration.* Power developed faster in the United States than in some other countries, so that its competitive position was improved. Americans have regarded this rapid growth with some pride, since it led for a while to stronger positions in money, culture, technology, standard of living, military influence, and so on.

Many properties of American life made this rate possible, and many of them became part of the values of the culture. In business there was the profit motive, the capitalist system, advertising, the sense of individual progress, and a general belief in the goodness of expansion. In other areas there was faith in the value of education, cluster living, a willingness to make cooperative efforts to overcome obstacles, a strong military that affected access to resources, and an especially heavy emphasis on good communication and transportation. All these things increased the ability to work cooperatively in bringing together various resources for the maximum rate of growth.

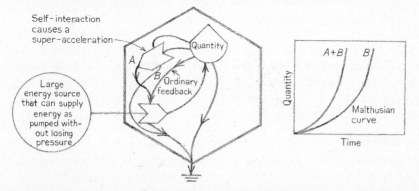

Figure 5-5 Model 4: Growth is faster with self-interaction pumping than with ordinary Malthusian feedback pumping. The *A* + *B* curve is the rate of growth in the United States in this century.

To represent the super-acceleration that develops in a competitive situation when sources of energy are rich, we have model 4, which includes a super-accelerating pathway. In Figure 5-5 we have modified the Malthusian growth diagram by the addition of pathway A to account for the special efforts that characterized the growth of the United States.

Self-interaction symbol.[3] The double-input block used in Figure 5-5 is a new symbol; we use it for any process which flows according to the amount of self-interaction.

For example, when people cooperate to produce a building, the total effect is not only their separate efforts but that which arises from their interactions. One person supplies information to another; one helps another hold pieces together while sewing or doing carpentry; one's enthusiasm affects another. Human activity is especially effective when there are good cooperative interactions. The double input is a faster action than a flow without self-interaction. The single pathway indicates that more action results from more people, but only in proportion to their numbers: two people do twice the work of one; three people three times as much. The self-interaction symbol indicates that two people do more than twice as much as one person, perhaps four times as much. But if they are accomplishing much more, they are also using up their reserves of energy for such work more quickly. The block symbol at A in Figure 5-5 has a double input showing accelerating self-interactions—one part of the society stimulating another. Note, however, that the special growth-stimulating work is twice as draining to the storage as ordinary pumping is. As long as the source of energy is large, the special acceleration efforts, although costing more drain from the storage, may accelerate more than they drain and thus result in faster growth. If, however, the source of energy is not large, such pathways lose their growth effect, becoming only fast drains.

Competitive Exclusion

The principle of competitive exclusion is well established in such situations as microbial cultures, competitions of weeds, and perhaps the economic competition of businesses. During periods of rapid growth, when sources of energy are relatively unlimited, and when no other controls are provided, one system accelerates its growth faster than another. The effect accumulates until one unit

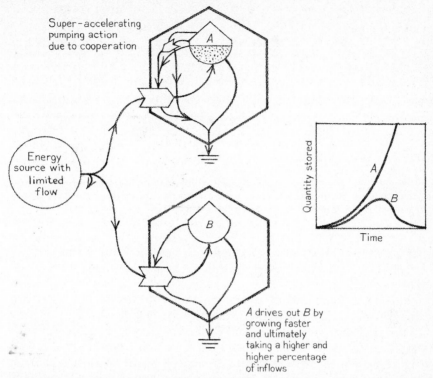

Figure 5-6 Competitive exclusion of one type of model by another. Model *A* has super-acceleration feedback that involves cooperation among the units of *A*. It wins. For example, suppose that people in *A* are cooperating and people in *B* are working by themselves.

drives the other out of business: one survives and the other does not. The survivor, by means of its accelerating growth, manages to capture resources for energy flows from the other.

In our culture the common belief in growth as necessary to survival probably comes from the principle of competitive exclusion. The principle is true during periods of expanding energy, and growth has been a characteristic of survival in the United States until now, when sources of energy are becoming limited.

Figure 5-6 shows competition between two growth units like those shown in Figures 5-3*c* and 5-5. The system with the greater feedback of resources into growth survives at the expense of the other.

MODEL 5: GROWTH WITH UNRENEWABLE SOURCE OF ENERGY

Figure 5-7 shows a system drawing its source of energy from an unrenewable source. There is no more energy than that in the initial storage: when the tank (*T*) is drained, there is no more. In this situation growth of quantity (*Q*)

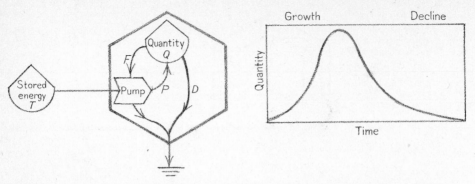

Figure 5-7 Model 5: Surge of growth followed by decline when energy source is temporary. Examples: the growth of beetles and fungi in a dead log as it decomposes; a mining village based on a body of ore that is used up, leaving a ghost town.

accelerates at first until the source (T) begins to run out; then the quantity of stored order (Q) gradually declines, as depreciation (D) and outflow of feedback (F) exceed production (P).

For example, a population of animals in a log eat the organic matter and the microorganisms and for a while a quantity of animals living on the log builds up; then the populations crest and decline as the source of energy runs out. Another example is a village running on hydroelectric power produced by a dam (Figure 2-2). If there were severe droughts, so that no more water ran into the dam, the village could grow for a while but eventually would decline as the water in the reservoir was exhausted. Many mining communities grew and declined as the ore was mined out.

MODEL 6: GROWTH WITH TWO SOURCES—ONE RENEWABLE, ONE UNRENEWABLE

Figure 5-8 shows a situation where energy is coming from two sources, of which one is temporary (a fixed initial storage) and the other is a regular renewable source. In this case, the sources of energy at first support a surge of growth; but the growth eventually declines back to a steady state based on the renewable source. The steady state is characterized by less development than the state which existed when both sources were contributing energy. The urban economy of humanity may be following this pattern, as we will consider in more detail in Part Three. Fish in a lake whose food chain is based partly on food dumped into the lake (an unrenewable source) and partly on food generated by plants using the regular and continuing inflow of light (a renewable source) are an example of model 6. At first there are two sources of food, and the fish population grows to a high level. But as the temporary food is used up, the fish population declines to a level that can be supported by the sun alone.

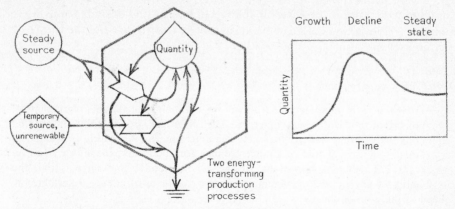

Figure 5-8 Model 6: Surge of growth and return to steady state when part of energy is temporary and comes from unrenewable storages. One example may be our national economy; another example is the ecological community in a new reservoir which uses the stored matter of water-covered soils.

GENERALIZING FROM DIAGRAMS

We have used the diagrams to generalize about systems of entirely different types that may be following the same patterns of growth. For example, the following seem different but all have similar graphs of increase over time: a culture of microorganisms, a population of deer, water in a tank, vegetation in a field, urban structure in a city. We have showed several main kinds of curves of growth and the energy-flow situations that produce them. Some of these involve nonliving systems, like water tanks; some involve living systems, like vegetation; and some involve the growth of the whole collection of living and nonliving structures that we call our human economy.

SUMMARY

Starting with the filling and discharge of a simple storage tank, we have examined how flows of energy cause growth, support maintenance, replace depreciation, and may cause competitive exclusion. From the nature and permanence of the sources of energy we can use models to predict the nature of growth, the existence of a steady state, and other patterns. Systems diagrams show when there will be growth, transition, steady state, or decline. Renewable force-maintaining sources generate extended, accelerating growth. Sources with renewable but limited flow sustain growth up to point of leveling off, after which a steady state occurs. Unrenewable sources can sustain only a period of growth, followed by decline. Human economy may now be following the graph that results when a renewable, limited flow is combined with an unrenewable flow. By means of overview diagrams we can formulate questions about reserves of energy and strategies for survival. And we can predict the future from the nature of predictable inflows of energy and designs for using it. (In Chapter 16 these kinds of growth curves will be used to predict the future of our economy.)

FOOTNOTES

1 The curves in this chapter were drawn by computer, using the mathematical equations that go with each diagram. The diagrams are actually differential equations with unspecified coefficients. The diagrams can be readily translated to equations by writing a rate of change balance for each tank: these are given on pages 271–272. What is important for the general reader to realize is that the availability of energy and arrangements for management of energy determine the shape of the growth curve.

2 When back pressure exists, our notation leaves the arrowhead off. For example, water pumped into a tank from the bottom flows against the pressure of water already in the tank. The inflow adds to the quantity in the tank only when the inflow pressure is greater than the back pressure. If the source is interrupted in such a situation, the flow can drain backward. Most pathways do not have much backforce, because they pass over energy barriers.

3 When a flow is accelerated by the number of interactions possible, it is in proportion to the square of that number. The pathway with a double-input block is a quadratic pathway.

Chapter 6

Net Energy
and Diversity

According to the maximum-power principle, surviving systems process energy flows in characteristic ways, because more power is obtained in these ways. High-quality energy is used to stimulate the processing of low-quality energy to higher-quality forms. Energy flows that are rich enough to produce more than they use up in this process can support other parts of the system that are not so rich. Energy flows that generate more high-quality energy than they use are said to produce *net energy.* Flows of net energy support parts of the system that are not rich enough in energy to run alone. When there is net energy, it can be used to stimulate more growth, to stimulate the inflow of energy from less rich sources, or in exchange for other energy flows.

Energy flows are usually interrelated because more power is produced when high-quality flows are made to interact in order to stimulate the low-quality flows. However, when everything is interconnected, it is often hard to see what parts of a system are rich and thus responsible for the vitality of the rest. Many decisions about energy are made incorrectly because the decision maker is not aware of the importance of hidden subsidies of energy from one part of a system to another. Because interacting energy flows are of different quality, ultimate energy costs are not easily calculated. In this chapter we use energy diagrams to trace the energy flows responsible for productive processes. With the help of a table showing the cost of converting one type of energy into another (Table 6-1, page 79), we can estimate the net energy of main energy sources and the amount of energy subsidy involved in other processes. The ratio of energy yielded to energy used (the *yield ratio*) is used to consider alternative decisions about the energy basis for humanity in these changing times. What is net energy? What are some of its best uses? And how do we measure "good use" of energy?

EVALUATING THE CHAIN OF INCREASING ENERGY QUALITY

In Chapter 2 (Figure 2-4), we introduced the idea of a chain of increasing energy quality: the low-quality energy of the sun develops food in plants, which develop wood, which is converted into coal and then through power plants into electricity. Low-quality energy is shown on the left, and the highest-quality energy is shown on the right after having been developed through successive stages of concentrating. Human activity is involved in the high-quality stages of the diagram toward the right. If we examine this chain in more detail, as is done in Figure 6-1, we find that each step also has a feedback, from right to left, of high-quality energy interacting to stimulate the processing of low-quality energy. We have already described how feedback interaction loops tend to develop because of the maximum-power principle (see Figure 3-3). Although the chain in Figure 6-1 is drawn as a line for the sake of simplicity, in the real world its shape is more like that of a web, with many sources and branches.

The feedback pathways exert control on the main energy flow from right to left. A few Calories of higher-quality energy have the ability to determine the time and place of work of a larger flow of low-quality energy. For example, high-quality energy in irrigation water, seeds, and human labor on a farm determines where the sun grows crops (see Figure 1-2). The ability to control a larger flow of lower-quality energy by interacting with it is what makes higher-quality energy valuable. As is shown in Figure 6-1c, the chain of transformations in quality, when viewed in space, converges and concentrates, whereas the feedbacks spread out. For example, the production of farms in an agrarian society converges from fields to farm to town to the wood-based power plant to the people (Figure 6-1b).

To generate the maximum amount of work possible, each type of energy must interact with an energy flow of another quality, as is shown in Figure 6-1a. Either it interacts (to the left) with a flow of lower-quality energy which it can control and upgrade or it interacts (to the right) with a higher-quality flow that will control it and upgrade it. High-quality energy is wasted if it is not amplified[1] in interaction with a larger flow of low-quality energy. Low-quality energy is underused if it does not attract to itself some high-quality energy for interaction. See Figure 6-2: more energy is produced in part *a,* where interaction of high-quality and low-quality energy takes place, than in part *b,* where such interaction does not take place.

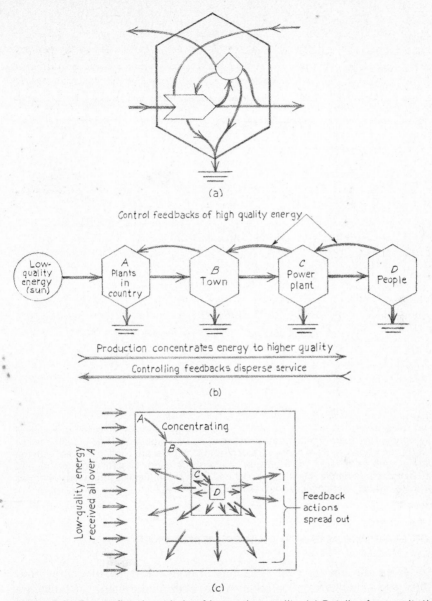

Figure 6-1 Energy flow in a chain of increasing quality. (a) Details of one unit. (b) Energy chain. (c) Spatial view showing high-quality energy and storages concentrated in a small area although they serve a larger area with feedbacks.

For example, electricity is a high-quality energy. If it is used to operate a dragline to process coal, it will generate more energy for heating houses than if the energy of the electricity is used directly for heating, as it is in resistance heaters.

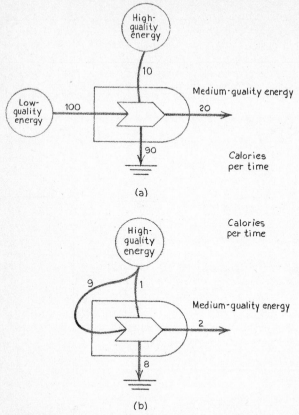

Figure 6-2 Comparison of the use of high-quality energy alone and as an amplifier of low-quality energy. (*a*) High-quality energy interacts as an amplifier of low-quality energy. Example: fertilizer, goods, and services produced from fossil fuel are used to interact with the sun in plant growth on a farm. (*b*) High-quality energy is used for both high-quality and low-quality purposes. Example: electricity produced from fossil fuel is used to make light as well as fertilizer, goods, and services for plant growth indoors.

TABLE OF ENERGY COSTS FOR TRANSFORMING ENERGY TYPES

Table 6-1 gives the energy costs of transforming one type of energy into another. To help us with our calculations of net energy, the energy of each type necessary to convert it to fossil fuel is given first. Our society uses many heat engines that run on fossil fuel. The table makes it possible to express other kinds of energy flows as equivalents of fossil fuel. For example, it takes about 2,000 Calories of sunlight to generate 1 Calorie of coal concentrated and ready for use. The second column in Table 6-1 gives the fossil-fuel equivalents of the type of energy listed, obtained by dividing the item in the first column into 1. Since it takes about 4 Calories of coal to generate 1 Calorie of electricity (including energy needed indirectly to run the plant), we show the fossil-fuel

Table 6-1   Energy Equivalents

Type of energy	Calorie cost (Calories of heat to make one FFE Calorie)	Fossil-fuel equivalents* (FFE Calories/per heat Calorie)
Heat from sun's rays, uncollected	10,000	0.0001
Sunlight	2,000	0.0005
Gross plant production	20	0.05
Wood, collected	2	0.5
Coal and oil, delivered for use	1	1.0
Energy in elevated water	0.33	3.
Electricity	0.25	4.
Money flow, 1970		$25,000 Calories/$

*Reciprocal of Calorie cost.

equivalent of 1 Calorie of electricity as 0.25 Calories. Where energy is written as Calories of fossil-fuel equivalents, we use the abbreviation FFE.

Since a few Calories of high-grade energy can have as much effect on work as many Calories of low-grade energy, we may compare effects by expressing various forms of energy as equivalents of one type. We have used Calories FFE for comparing the energy costs of different flows.

NET ENERGY

When we consider the declining reserves of fossil fuel in the 1970s, the question whether energy-transforming activities generate net energy becomes very important. All processes have energy feedbacks, and it is very easy to forget about the return pathways of energy subsidies. If these energy flows are high-quality work, such as the work of human beings and complex machines, we may forget how much energy it took to develop them in work elsewhere in the economy. Much of our thinking about energy is wrong because feedbacks of goods and services have not been considered on a comparable basis.

Net Energy of a Source

Net energy of a source is the energy yielded over and beyond all the energy used in processing it. Figure 6-3 illustrates this concept. Energy from oil reserves (source) is being processed by higher-quality energy feeding back from elsewhere in the economy. The energy required for pumping oil and for maintaining the machinery, assets, and personnel of the oil company (energy flow *A*) is less than the energy yielded (energy flow *B*). In recent world history, large amounts of net energy from oil and coal (fossil fuels) have been used to support other activities that could not operate alone without this subsidy.

In contrast, a source of energy which does not yield net energy is shown in Figure 6-4*a*. Energy flow *A* is larger than energy flow *B*. An example of such a

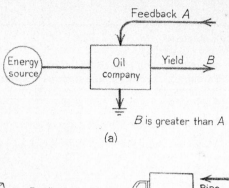

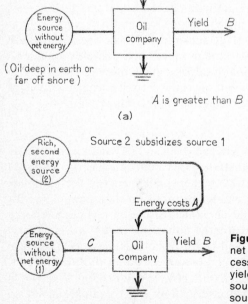

Figure 6-3   Energy transformation with net energy: yield *B* is greater than feedback *A*. The energy source does yield net energy beyond the energy required for its processing.

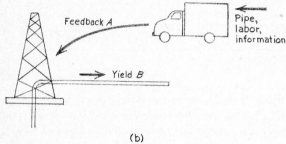

Figure 6-4   (*a*) Energy source without net energy. The energy needed for processing (*A*) is greater than the energy yielded (*B*). (*b*) Use of a second energy source (2) to generate energy from source 1, which does not yield net energy on its own.

source is oil deep in the ground and far out to sea: the cost of building steel towers, operating pipelines and boats, repairing corrosion and damage from storms, and operating all the complex services required for such an elaborate system may be greater than the energy yielded.

Subsidizing One Source with Another

Although the process in Figure 6-4*a* does not yield net energy, energy from another source may be used so that the process can go: see Figure 6-4*b*. Both sources contribute some energy, and the second one contributes some of its excess (net energy) to make the first yield energy. The first source still has no net energy, but the first and second together do generate net energy. According to the maximum-power principle, this may be a good use of energy, since the excess energy of the rich source (2) is not wasted but is used to increase the energy flow.

Evaluating Net Energy

Net energy may be evaluated by means of our energy diagrams. Since interacting energy flows are of different quality and thus represent different amounts of energy ultimately involved, we must estimate the energies used by expressing the energy flows in fossil-fuel equivalents (FFEs). This is done in Figures 6-5 through 6-7. First, in part *a* of each figure, ordinary Calories of heat equivalents are written on the pathways. At each point in these diagrams, inflowing Calories are equal to outflowing Calories—as must be true according to the first principle of energy.

Next, in part *b* of each figure, fossil-fuel equivalents are estimated and written in. In Figures 6-5 and 6-6, the yield and the flow from source 2 are already in the form of concentrated fossil fuels. Therefore, the fossil-fuel equivalents are numerically equal to the heat equivalents.

In Figure 6-7 the energy flows are associated with generating electricity from photoelectric cells of the type used to power satellites and used in light meters. Some people have advocated converting solar energy directly to electricity by means of such devices. Figure 6-7*a* shows the energy flows for 1 square meter of such cells each day. It includes the average daily cost of such cells, including the cost of replacing them every ten years. Energy flows were converted from heat equivalents into fossil-fuel equivalents, using Table 6-1, to produce the numbers in Figure 6-7*b*. The result is no net energy. Much fossil-fuel energy is spent through the economy to maintain the cells, with a very tiny return.

To help us visualize net energy, Figure 6-5*c* and Figure 6-6*c* show the output energy turned back to supply the high-quality fossil-fuel energy. In Figure 6-5, this is possible, and 10 Calories remain as net energy. But in Figure 6-6, the feedback is not enough to supply the 400 Calories required for the process: there is no net energy. The subsidy is the 400 Calories minus 90 Calories—that is, 310 Calories. The process does yield energy. But we must ask: Is this as high a yield as that of some other processes that could be subsidized?

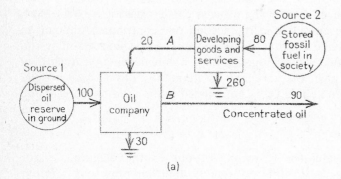

(a)

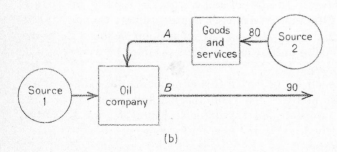

(b)

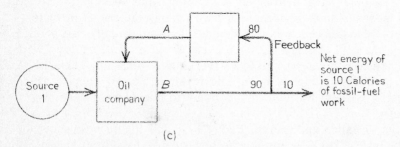

(c)

Figure 6-5 Example of a rich energy source with net energy. Oil company gets net production of energy from pumping and processing oil. (*a*) Energy flows are Calories of heat equivalents per day. Solar energy is in excess of whatever is needed for fossil fuel to develop its maximum work per day. (*b*) Fossil-fuel equivalents. (*c*) To test for net energy of source 1, substitute feedback of outflow oil for source 2.

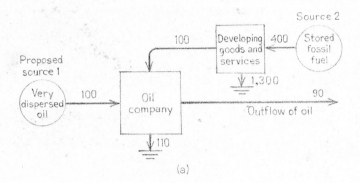

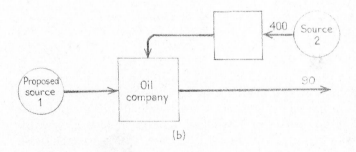

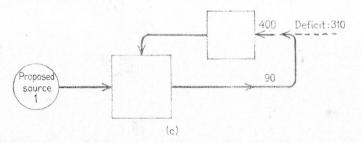

Figure 6-6 Example of a poor energy source, which does not yield net energy alone. Energy from a second source is required to deliver energy from the first source. Solar energy interacts as necessary to maximize the work of the high-quality source (2). (*a*) Energy flows are Calories of heat equivalents per day. (*b*) Fossil-fuel equivalent Calories per day. (*c*) To test for net energy of source 1, substitute feedback of outflow oil for source 2.

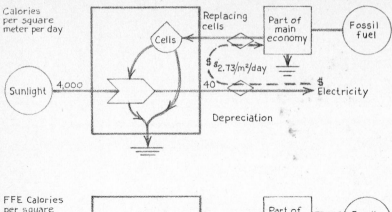

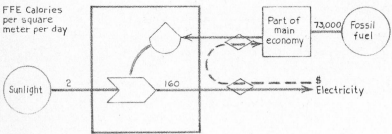

Figure 6-7 Energy flows in generating electricity from silicon photocells 1 square meter in area. (*a*) Calories of heat equivalents flowing per day. (*b*) Fossil-fuel equivalent (FFE) Calories flowing per day. There is no net energy, and the yield ratio is very low.

In Figure 6-5*a* high-quality fossil-fuel energy is shown interacting as may be necessary with low-quality solar energy to generate goods and services that our economy supplies to the oil companies. This again illustrates the principle that high-quality energy interacts with low-quality energy in generating work that fulfills the potential of a system. Systems which do not increase their work in this way do not generate as much work and will therefore not be competitive.

Consider the yield of energy from buying Arab oil, shown in Figure 6-8. In

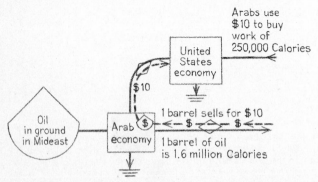

Figure 6-8 In 1974, the United States received fossil-fuel equivalents from the Arabs in the ratio of 6.4 FFEs received to 1 FFE invested in products sold to the Arabs.

NINE TO FIVE by HOLLAND

" One's oil and the other's dollars."

Holland—London Sunday Telegraph (Rothco).

1974, 1 barrel of oil from Arab sources cost $10. If $1 represents 25,000 Calories in FFEs (see Chapter 4), then paying $10 for 1 barrel of oil is equivalent to sending 250,000 FFEs to the Arab countries in exchange for it. Since 1 barrel of oil yields about 1.6 million Calories, the yield ratio is therefore 1.6 million divided by 250,000, or about 6.4. This figure is very high. In Chapter 11 we will find that most sources of coal and other energy remaining in the United States have ratios lower than 6. With such a high return for energy invested, it is easy to see why there has been a tendency for all the world to become dependent on these rich sources of oil, and to remain so even after the large increase in price of 1973.

Energy Yield Ratio

In order to evaluate sources of energy to be used by a society, it may not be sufficient to determine simply if there is net energy. Sources which yield some net energy still may not be able to compete if they yield less than other sources. For comparing sources, an *energy yield ratio* can be calculated. This is the ratio of energy yielded to energy fed back from high-quality sources, with both terms expressed in fossil-fuel equivalents. In Figure 6-4b the yield ratio for energy source 2 is the ratio of B to A. For source 1, the yield ratio is the ratio of B to C.

When the value of the yield ratio is greater than 1, there is net energy. In Figure 6-5 the ratio is 90:80 (1.1), a small net energy. In Figure 6-6 the ratio is 0.23, less than 1, without net energy. In Figure 6-7 the ratio is only 0.002. In Figure 6-8 oil from the Near East is a large net yielder even at $10 a barrel, with a yield ratio of 6.4.

STRATEGIES FOR INVESTING NET ENERGY

When net energy is produced, this means that more high-quality energy is being generated than is fed back as energy costs of a process. If a system is generating net energy, what can it do to maximize its chances of survival? The energy processed and transformed (1) can be stored as growth of the same kind already present, accelerating the original process; (2) can be used to diversify and develop additional sources of energy or more efficient processing; or (3) can be exchanged as yield in order to obtain more special imports from outside. According to the third principle of energy (the maximum-power principle), the alternatives that generate the most energy are the ones that will be successful and will foster the survival of the system in competition for energy. These three alternatives are shown in Figure 6-9. If the existing source of energy is still large and relatively unused, more use is possible. Then, *continued growth* of the same kind will feed back more rapid use of the present source. If there are additional sources of energy within the system that are not well used for survival, or if the processing of energy is inefficient because there are not enough specialists in some field of work, then more energy is achieved by building *diversity*. If there are rich sources of energy in a neighboring system which can be obtained by trade, *export* may be the best use of net energy: more energy may be obtained by sending out the energy as yield to start a new exchange. Most systems do some of each. Examples of each of the three strategies follow.

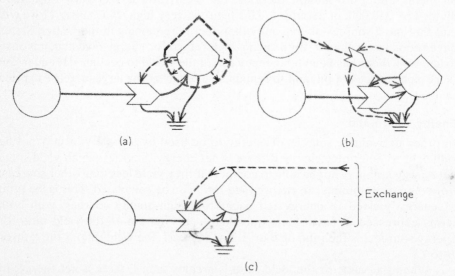

(a)   (b)

(c)

Figure 6-9   Three choices for using extra stored energy to get more energy. The dashed lines indicate new pathways. (*a*) Use the present source faster, as in simple growth. (*b*) Start tapping a second source—diversity. (*c*) Trade with outsiders for additional resources—exchange.

   1 New agriculture in America used net gains to produce more of the same kind of development because there was more land of the same type to use close at hand.

   2 Some energy stored from agriculture was used to build water wheels which harnessed the energy of streams as a second source.

   3 Some of the energy from agriculture and water mills was put into products for exchange (by barter or money purchase) to bring in additional energy such as oil.

Agriculture—An Example of Subsidy to Increase Yield

Many systems have their yield increased by the addition of energy from a second, rich source to a process that had some net energy but only a relatively small amount. In intensive modern agriculture yields are increased by feedback of special work of goods and services that ultimately are derived from fossil-fuel support from the main economy. Figure 6-10 shows the simple agriculture done by primitive peoples with no subsidy of outside energy and compares this with intensive modern agriculture, which is heavily subsidized by flows of energy to make fertilizer, pesticides, machinery, special varieties of seed, fuels, etc. Simple agriculture was tested under severe conditions of survival over thousands of years and developed a net yield of food energy as the basis of human life before the industrial revolution. This use of dilute solar energy by human beings probably represented the maximum possible without

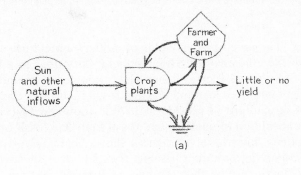

(a)

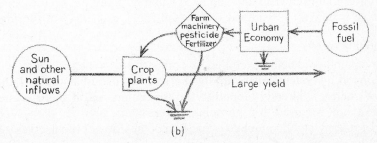

(b)

Figure 6-10 Energy basis for agriculture. Subsistence agriculture (*a*) has smaller yields than subsidized, modern agriculture (*b*).

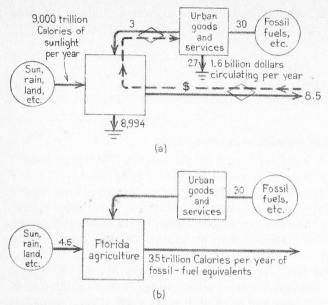

Figure 6-11   Main energy inflows to Florida agriculture on 15 million acres. Conversion of sunlight to food is heavily subsidized by a second energy source, fossil fuel, that supports machinery, fertilizer, pesticides, the standard of living of human laborers, etc. (*a*) Data for examining Florida agriculture. Numbers are trillions of Calories of heat equivalents per year. (*b*) Fossil-fuel equivalents of flows.

special extra subsidies. The much higher yields of modern agriculture are due not to any increase in the efficiency of using sunlight, but rather to a hidden subsidy of the process by fossil fuels. Higher yields allowed less land to be used; as a result, sunlight was used less and less and fossil fuels more and more.

The condition of agriculture in Florida in 1970 is shown in Figure 6-11, which indicates the contribution of sunlight and fossil fuel. The fossil fuel was estimated from the money flows involved. Part *a* shows heat equivalents and data for the flow of money in agricultural sales. Part *b* shows the fossil-fuel equivalents for these flows. Notice that the energy from fossil fuels (30) is much greater than that from the sun (4.5) when both are put on the same basis. Agriculture is heavily subsidized with fossil fuels, especially in Florida.

Developing Diversity

To survive—to avoid being excluded by competitors—when energy is available for expansion, a system must rapidly increase its use of energy. If the main source of energy is already tapped to the maximum, and if there is energy of production not required for pumping the primary source, then secondary sources of energy can be pumped even though they do not themselves yield net energy. For example, our pioneer agricultural economy used windmills. Since catching the wind was not possible without the wooden structures that were used for early windmills, wood as a source of energy was being used to subsidize the obtaining of additional energy from the wind. Windmills might not have been possible without the second source of energy (wood from the sun).

Sometimes a secondary source replaces the primary one. For example, an early agricultural economy in America, which heated its houses with wood, began to dig for a second source of domestic heating, coal. Soon coal became the major source and wood the secondary source. Primitive hunting cultures developed agriculture on the side; but agriculture later became the main source of their support and hunting a minor source.

Having diverse units enhances the use of energy and the probing for energy. Notice in Figure 6-12 how energy is first stored and then used to support diversity that feeds energy back to tap other sources. The development of a high degree of diversity can favor the collection of energy and the survival of a whole system. Development of diversity also provides flexibility in case there are changes in the relative availability of energy resources. A diversity of occupations in a city tends to make an efficient economy, since there is a specialist who does well each job that needs doing.

For example, a coral reef has a great many different kinds of fish. Each species of fish, in order to recognize its mates and to mark its territories, has developed a unique color pattern. Making color requires chemical energy, and energy is also required for the recognition of colors and the response to them by sense organs and brain. If there were only one kind of fish, this would not be needed; diversity, then, requires energy. The diversity of fish feeds back services to make the whole fish system more efficient. Each kind of fish consumes specific organisms; this creates a complicated food web. Such interdependence means that all parts of the reef can be used.

The energy required for organizing diversity is large, since much coordinating and processing of information is required where there is specialization and division of labor. Notice the many heat-flow losses in Figure 5-2a. Diversity can be both an aid to energy and a drain on energy. Notice the double-input block in Figure 6-12b, like the one in Figure 5-2; it shows the special, high energy cost of maintaining diversity. Main storages of energy are used to provide diversity, and the diversity feeds back to stimulate other, minor sources of energy. The ultimate effect is that more energy is processed; the system has more variety; and the system is more flexible if there is an interruption to its first main source of energy.

There are many examples of the situation shown in Figure 6-12. Let us consider a human example, where the main source of energy is fossil fuel and other sources are winds, waters, tides, and geological movements. Suppose the main source supports a city's major structures, including the major human needs for energy. Some of the main energy is fed through specialists who interact with other sources to obtain additional inflows to this system. For instance, specialists process water, use winds to air-condition, and develop tourism on beaches that receive waves. In Figure 6-12, A, B, and C are the separate packages of information that go with the special jobs.

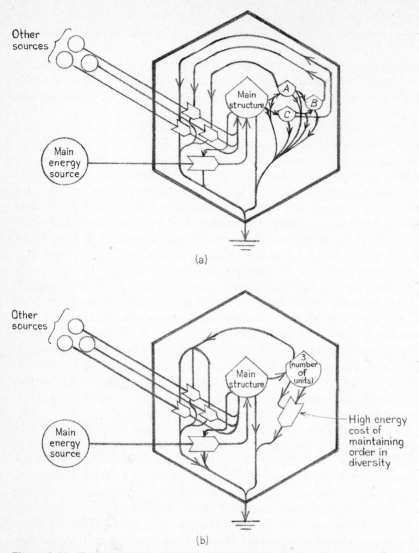

Figure 6-12 Two ways of showing diversity of units in a system that uses net energy from the main energy source to obtain energy from other sources. (a) Diversity units (A, B, and C) are shown separately. (b) Diversity units are combined as one storage.

Now consider an ecological example when the main source of energy is the sun and the main structure is the biomass of a forest. *Diversity* here refers to the special information in a variety of species. Other sources of energy are the flows of water and wind, organic matter, and geological input. The specialists help incorporate the other flows of energy into the main system of energy. For instance, some species of trees with long roots may reach deeper for mineral nutrients from the rocks; flowers may use the winds in pollination; leaves use the dry wind to evaporate their waters from microscopic leaf pores for cooling. People who study ecological systems such as forests and estuaries believe that diversity of organisms works in this way to make such natural areas better able to survive in varying conditions. Many people who study human economic systems believe that a varied economy is a more stable one.

Exported Yield versus Local Diversity

Often in agriculture, in forestry, and in general economic planning there is a choice to be made whether a local system should be self-contained and develop more of its own diversity and variety, or whether it should send more to the external world as trade goods and services for exchange.

For example, most countries in precolonial times had local economies which exported for trade relatively little of their energy. Their economies were diversified enough to supply their own local needs. As rich resources began to be found and applied to rail and ship transportation, the situation changed. Local areas developed a cash crop or some other specialty which they exported to get funds to buy fossil fuels or the products of fossil-fuel economies. In the colonial period many communities changed from locally based economies to export economies. The rising levels of energy throughout the world changed the norm from local diversity of occupations to diversity of interacting countries each with a specialty.

Another example is the pattern of forest management. At one time, the best course was to let a forest operate and maintain itself. The forest used diversity of species as one means of preventing ravages by insects and epidemics of disease, controlling fire, etc. Later, when rich fossil fuels became available, the more economic method of management was to develop a monoculture (one kind of a tree) and sell wood for the money to buy goods and services based on fossil fuels to fertilize, weed, poison, and harvest. Now that fossil fuels are becoming less available, the best course shifts back to creating a more self-contained forest with more diversity.

SUMMARY

In this chapter we have considered situations that yield net energy for growth and the choice of pathways that a surviving system will select from: to accelerate the growth of more of the same, develop more exchange for energy from outside, or add local diversity to tap all local energy most efficiently. Most systems do some of each.

Diagrams were used to try to untangle confusion over such terms as *net energy production* and *net growth*. *Net energy* was defined as the yield in excess of the cost of feedback energy. To be accurate, we must express all flows of energy in the same units of ability to stimulate work. Calculating net energy requires converting various forms of energy to fossil-fuel equivalents (FFEs). A flow of energy can be used to subsidize a second flow, so that energy can be obtained from a source which yields no net energy of its own. To decide which energy flows are good ones in which to invest rich forms of energy, we use the energy yield ratio, defined as the ratio of energy yielded in FFEs to the energy invested, also in FFEs. Systems grow and diversify where yield ratios are high. Very high yields are available from Arab oil, even at high prices. In Chapter 7 we will consider how alternatives for investing energy are involved in natural ecological systems such as forests, lakes, and seas.

FOOTNOTES

1   The interaction process and the symbol we use for it can also be called an *amplifier*. An amplifier is a structure with two energy inflows that interact to produce an energy output. The action of one energy inflow, usually a varying one, increases the output of the amplifier proportionately, using the second flow as an additional source of energy.

    In electronics the inflow that is amplified is called the *signal* and the second source of energy is called the *power supply*. An example is a voice amplifier; the variations in tones of the voice are the signal and the power is supplied from the electrical outlet. In economic discussions a small effect is called an amplifier when it interacts with another flow to cause a larger effect.

    Amplification implies flows of two qualities of energy. The low-quality flow supplies a large quantity of Calories; the high-quality flow supplies the controlling action with a small quantity of Calories. The interaction stimulates the flow of both. Although the high-quality flow is said to be amplified in electronics, it is probably correct in other areas to say that either or both are amplified by the interaction.

Chapter 7

Flows of Energy in Ecological Systems

Ecological systems illustrate the principles governing flows of energy. They have been on earth for millions of years, and have adapted over long periods to maintain their species, pass their heritage on to the future, and maintain a cover of vegetation over the land and aquatic life in rivers, lakes, and seas. This chapter gives some examples of ecological systems to show how such systems process energy and build order. Because these natural systems are ancient and have adapted their species over a long time, they may help us learn something about the kinds of systems that follow from various energy conditions. We are able to derive more principles about all systems from examining the systems of nature.

There are ecosystems in water and on land, in dry and wet regions, in cold and hot regions. All of them seem different, with different species of plants and animals and different adaptations to special climatic and geological conditions. However, all have common aspects that follow from similarities in the basic flow of energy that drives all systems. Energy diagrams help us compare ecosystems and recognize equivalent roles in different systems. Some ecosystems are rich in energy and full of life. Others are poor and unproductive. Some are new, changing, or growing, and some are old and in a steady state. Some ecosystems store some organic matter or yield it as an export; others use more organic fuels than they produce—the latter gradually run down, unless they have an outside source of energy.

PRODUCERS AND CONSUMERS

Scattered over the land and in the seas are ecological systems—called *ecosystems* for short—in great variety. They contain plants, animals, and microorganisms which process energy, maintain structures, sustain cycles, and operate according to the laws of energy we have already discussed. Examples include forests, tundra, grasslands, eutrophic lakes, kelp beds, coral reefs, blue-water tropical seas, and swamps. All these various ecosystems have some common characteristics that follow from the principles of energy. There is a basic plan common to all the solar-based ecosystems. They all have a cover of plants that use sunlight to make food (the process of photosynthesis); we call these plants *producers*. These systems also have varied groups of organisms, including both plants and animals, that use the food; these are appropriately called *consumers*. They also perform services for the system, maintain themselves, and recycle materials.

Symbiosis of Producers and Consumers

Figure 7-1 shows the basic pattern of most ecological systems. Part *a* gives the patterns of energy flows, the productive upgrading of energy, consumption, and some recycling of energy and its dispersal into the environment of the system. Part *b* presents the overall processes in a balanced aquarium such as many readers may have assembled at school or at home.

As Figure 7-1*b* shows, light coming through the glass stimulates and supports the photosynthesic production of organic matter that serves as food for all the living parts of the system. The term *organic matter* refers to carbohydrates, fats, proteins, wood, peat, skeletons, and all other components of the bodies of organisms, living and dead. Organic matter accumulates in storages in plant matter; living at first, and later as dead matter such as logs and lake muck. Heat disperses in the environment, and oxygen is produced. Meanwhile, the consumers of the aquarium are eating the organic food matter directly or indirectly and running their bodies on this food. To consume and use the food, they breathe in oxygen from the air and use it to combine with the food matter. The by-product of the consumers' action is a breathing out of carbon dioxide, water vapor, mineral nutrients, and more dispersed heat. *Mineral nutrients* are all the inorganic chemical substances that are required by life and taken up by plants from soil and water. They include the fertilizer elements phosphorus, nitrogen, potassium, calcium, magnesium, sulphur, and many others.

Another name for the consumer process is widely used in biology: *respiration.* This is what people do, for example, when they are breathing and

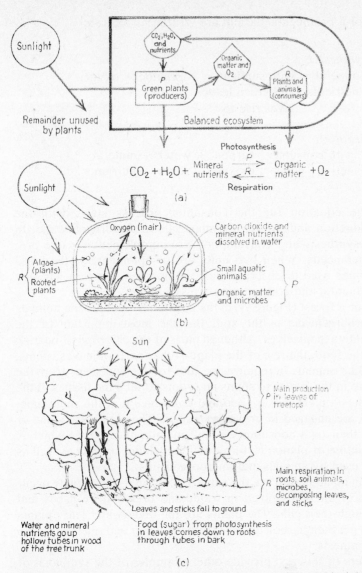

Figure 7-1 Energy flows of the production-consumption symbiosis. (*a*) Energy diagram. (*b*) A balanced aquarium. (*c*) Forest ecosystem showing location of production and respiration. Note main pathways of water and nutrients going up and organic matter coming down.

living. The term can also be used for the "breathing" of a car or a city. The overall reactions are the same. Consumers, especially microbial consumers, decompose their organic food into mineral nutrients and are therefore sometimes called *decomposers*.

The reactions of a balanced aquarium are written as follows:

Production (photosynthesis):
Carbon + water + mineral   with sunlight   organic + oxygen
dioxide   nutrients   and green leaves   matter
                                   gives rise to

Consumption (respiration):
Oxygen + organic   with consumer   carbon + water + mineral
        matter   activity gives   dioxide   vapor   nutrients

Notice that the equation for the consumer process starts with the by-products of production and returns the materials to the original state. In other words, photosynthesis and consumption are complementary: one provides what the other needs. When two units are complementary, the word *symbiosis* is sometimes used to describe their relationship. We may say that photosynthetic production and consumption are symbiotic.

In a terrarium or a forest (Figure 7-1c) the consumers include the animals and the many microorganisms in the soil. But the most important of the consumers are the plants themselves. Although parts of the chlorophyll-bearing green leaves make the food, the rest of the plant—other parts of leaves, stems, roots, etc.—is just like animals in requiring regular food. Food seeps from the place of manufacture in the leaves to stems and roots to feed the tissues of the rest of the plant. Thus a plant makes food and consumes it at the same time. Plants making food are engaged in *gross production*. The plants use some of this production for their own consumption (self-maintenance). That which is left is the *net production* of plants. Net production includes oxygen as well as plant tissues and is used by animals and microorganisms.

The processes in an aquarium are the same, except that they mostly occur under water. The oxygen is dissolved in water and breathed by fishes and aquatic insects through their gills. The plants are the underwater leafy plants and the algae which get their carbon dioxide from the water and release oxygen to it during photosynthesis.

Previous chapters of this text provide other examples of the symbiosis of producers and consumers. The first example, the farm in Chapter 1, shows plant production, except there we do not keep track of the gases, oxygen and carbon dioxide. In that example, these gases blow in or out with the winds and are relatively unlimited. The food was consumed externally.

Figure 2-3 shows the reaction of a forest fire in which oxygen and wood burn to produce carbon dioxide and water vapor. This is the same reaction as that taking place in living consumers. A fire is another kind of consumer. As its energy storage, it makes turbulent eddies in the atmosphere as the smoke billows and rises. We use fires in boilers to run power plants and cities; we use explosive combustion in cars.

Human beings, too, are consumers and have the same overall respiration reaction as that given in Figure 7-1a. An individual is personally a consumer as he eats food, breathes in oxygen, and breathes out carbon dioxide and water vapor. Our machines are also large consumers, using the same raw materials and releasing the same waste products. A city as a whole is a giant consumer, like a big animal.

If farms are examples of photosynthetic production, and villages and cities are examples of consumers, then the combination of farm and village makes a symbiotic pair which can be described by the same general diagram that is shown in Figure 7-1a. Figure 3-4 shows a cycle of materials circulating between farm and village. From Figure 4-1 we learn that chemical substances circulating between them are examples of the production-consumption cycle.

The greatest example of the balanced symbiosis of producers and consumers is the whole biosphere. The plants of land and seas are making food and oxygen, and the consumer parts of plants, soils, marine organisms, and cities are making carbon dioxide and water vapor to be used again.

Mineral Cycles

In Chapter 3 we showed that there is a cycle of chemical raw materials between plants and consumers. Now that we have shown the production process (photosynthesis) and the consumer process (respiration), we can identify some of the main chemical substances that make the circle from plants to consumers and back to plants again. We call the circulation of these materials the *mineral cycles*, or sometimes the *nutrient cycles* because the circulating elements are required for the nutrition of the plants and animals. The circle they make is shown in Figure 7-2. Compare Figure 7-2 with Figure 7-1. The dotted pathway diagrams are the same where energy and matter are traveling together. The circulation of matter as part of a system of energy flows of the production-respiration ("P-R") pattern was introduced in Figure 3-4 as an example of the general principle that flows of energy develop cycles of materials.

The largest mineral cycle is the carbon cycle. Carbon in carbon dioxide is utilized in the plant production process and incorporated in organic food, as part of which it is passed to consumers. The consumers turn carbon back

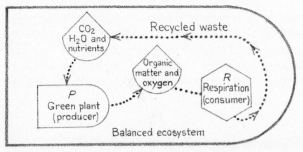

Figure 7-2 Mineral cycle in ecosystems that are closed to matter. Flow goes along the dotted path from plant producers and the production process (*P*) to storage of organic matter to consumer respiration (*R*) and then to storage as mineral nutrients ready for photosynthesis again. The energy flows for this situation are shown in Figure 7-1.

into carbon dioxide, completing the cycle. Notice the closed pathway in Figure 7-2.

Another main nutrient requirement is the element phosphorus. Phosphorus is present in soils and waters in a dissolved chemical state called *phosphate.* It may have originated from erosion of rocks, or it may have entered in small amounts from the rain. Plants incorporate phosphate as needed for their biological porcesses. Phosphorus is passed to consumers in organic matter. The consumers use the organic matter for food, and the phosphorus is released again as a waste product; for example, wastes from the kidneys of animals are rich in phosphorus. Farmers add phosphorus fertilizer to crops to increase growth.

Nitrogen is another important nutrient requirement. Plants get nitrogen from nitrogen salts dissolved either in the soil or in water and build it into the protein structures of the organic matter they produce. The nitrogen, in organic combination, is passed on to consumers in the organic matter that they eat. After the food is consumed, the nitrogen is released through the animal kidneys or as nitrogen salts in other wastes. Nitrogen salts reaching the plants are ready for use again. Nitrogen takes different chemical forms in various alternative pathways in different situations: ammonium salts, nitrites, nitrates, organic amino compounds, and many others. For our purposes we need show only the overall main cycle; see the dotted line in Figure 7-2.

Four-fifths of the air in our atmosphere is nitrogen, and this can be incorporated into the nutrient cycle if a system has the special structures and the extra energy necessary to convert it from a gas to a more easily used salt form. Many plants have nodules on their roots for fixing nitrogen from the air into a chemical salt form. Industrialized agriculture uses fossil fuels to operate industries that fix nitrogen into fertilizers. The lightning in thunderstorms also converts some of the nitrogen gas in the air into nitrates. In all these examples, considerable energy is diverted from other work to change nitrogen to usable forms. Where there are already enough nitrogen salts circulating, the special process of fixing nitrogen from the air is not necessary, and the energy cost is less.

Visualizing these three chemical cycles—carbon, nitrogen, and phosphorus—may be easier if we indicate their relative magnitudes in most ecological systems. About half of organic matter is carbon. The ratio of carbon to nitrogen to phosphorus in organic matter is about 100:16:1. Even the water that is used by plants to make organic matter and released again by consumers as water vapor or through their kidneys is part of the chemical cycle: water is made of hydrogen and oxygen, and both of these are part of the cycle as given in Figure 7-1.

"P-R" Diagrams

The production-respiration symbiosis may be diagramed in various degrees of detail (Figures 7-1 to 7-8). The production process—photosynthesis—is labeled "P"; the process of respiration by consumers is labeled "R." Sometimes we call this overall symbiotic process the *balance of P and R.*

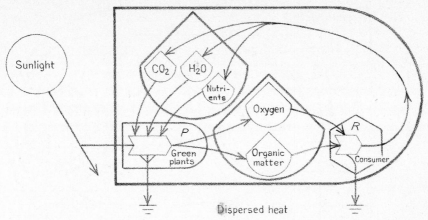

Figure 7-3 Producer (*P*)–consumer (*R*) symbiosis of ecological systems. Chemical components are isolated to show how they cycle in the *P-R* process. Examples: balanced aquarium,
pond, mountain forest.

In diagraming the overall processes, we can show more or less detail
depending on whether we separate out the various raw materials and by-
products or leave them in a group. Figure 7-1 is a simple arrangement with the
parts of the reactions clustered in two storages and two processes. In Figures
7-2, 7-3, and 7-4, the reactants are separated out. Figure 7-2 shows the mineral
cycle but ignores energy. Figures 7-3 and 7-4 include energy pathways that
accompany chemical components when they are shown separately. Figure 7-4
differs from Figure 7-3 by including the storage of living structures for plants
and for consumers.

The reader may still wonder why the pathways that are labeled according
to the materials flowing—e.g., phosphorus, carbon, organic matter—are lines
on an energy diagram. The reason is that there is an energy component and
contribution for each of the flows required for any process. Each of the
materials contributes a portion of the energy requirement. Some of the flows
may be high-quality energy and others may be low-quality energy. In the
diagrams the by-products include the heat released, as shown by heat-sink
arrows, and the waste nutrients, which emerge at the same time. The nutrients
are shown coming out with the heat but going on separately to be cycled and
reused.

Food Chains among Consumers

In Figure 7-4 there is a three-stage pathway through which solar energy passes
in developing storages of organic matter. First, sunlight develops organic
matter in plant tissues; second, the organic matter goes into a dead form such
as wood and forest litter; third, some of the organic matter is processed to
become the organic matter in the bodies of consumers, such as in the tissues of
insects eating the wood. The three stages constitute a chain of food processing
in which some energy is lost at each stage. The quality of some of the energy is
upgraded at each stage, as is described in Chapter 2 (see Figure 2-4). Figure 7-4

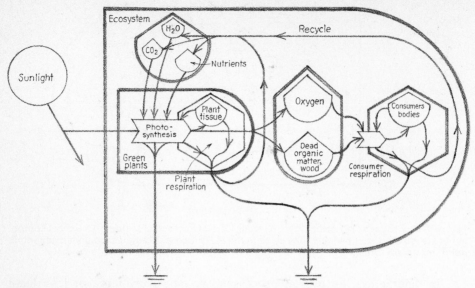

Figure 7-4   Producer-consumer symbiosis; here, storage of organic matter is shown in three places, as in nature (plant tissue, dead organic matter, consumers). Many pathways are still not shown. Examples: balanced aquarium, pond, mountain forest.

shows more detail about the relationship between producer and consumer, but the basic P-R symbiosis is still there, and the simpler diagram shown in Figure 7-1 is still a correct summary.

Most ecological systems develop even longer food chains, with at least five stages. For example, in a fishpond algae support small water fleas (plankton), and these support minnows. The minnows support perch, and the perch support bass. However, not all the food goes directly through the fish food chain. Part of it goes first into dead organic matter, often as particles in water or in the mud at the bottom. Living among these particles are microorganisms (microbes) which consume the organic matter. Animals eating the organic particles are also eating microbes, which are very nutritious. The food pathways are usually branched, with each species eating from more than one pathway; the result is sometimes called a *food web*. A fairly typical food web is shown in Figure 7-5. The overall P-R pattern is still there, but the consumers are diagramed in greater detail to show the main parts of the ecosystem. Notice that, as was already mentioned, parts of plants are consumers, utilizing food.

Ecosystems with Inflows and Outflows of Matter

In order to emphasize the production, respiration, and mineral cycles of the typical ecological system, the diagrams and examples given so far in this chapter have shown a closed cycle of matter without either organic matter or mineral nutrients entering or leaving a system. Only sunlight was shown coming into the ecosystem and only dispersed heat going out. Most ecosys-

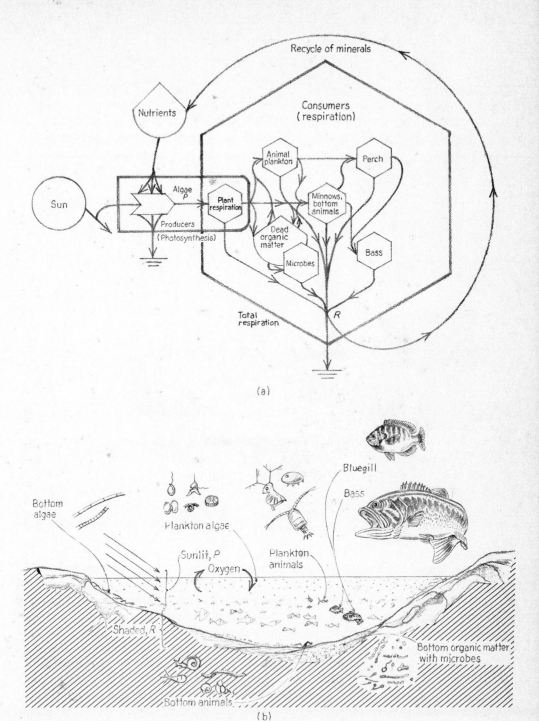

Figure 7-5 Producers and consumers of a pond; details of the food web are shown. (a) Energy diagram. (b) Pond; location of the main organisms of the food chain are shown.

"The devil with the food chain. I *like* mercury."
Sidney Harris, The American Scientist, September–October 1971.

tems, however, have inflows and outflows of matter. Three examples are given in Figures 7-6, 7-7, and 7-8.

Figure 7-6 has an inflow of organic matter that stimulates more respiration than production, and the outflow of matter is in the form of the mineral nutrients, a by-product of respiratory decomposition. A sewage plant is an example of this kind of system. After chunks of solids are settled out, water containing much dissolved organic matter flows through a secondary treatment that is really a domesticated ecosystem. The trickling filter (Figure 7-6b) houses a variety of small microscopic animals and bacteria that decompose the organic matter as it goes through.

Figure 7-7 has an inflow of inorganic mineral nutrients which stimulates photosynthetic production so that production exceeds respiration and there is a net production of organic matter. The example given in Figure 7-7b is an

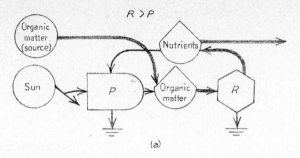

(a)

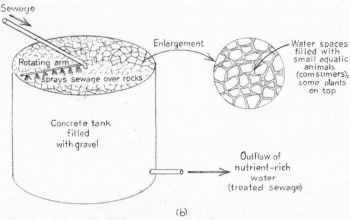

(b)

Figure 7-6 A steady state with an inflow of organic matter and an outflow of nutrients. (*a*) Energy diagram. (*b*) Example: trickling-filter ecosystem for treating sewage (*treating* means *decomposing organic matter*).

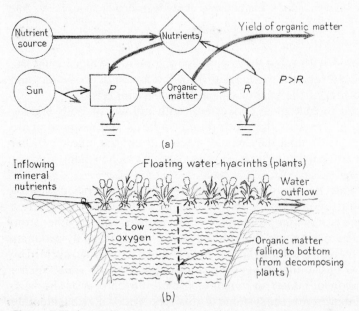

(a)

(b)

Figure 7-7 A steady state with an inflow of nutrient and an outflow of organic matter. (*a*) Energy diagram. Example: eutrophic lake receiving nutrients of treated sewage. (*b*) Eutrophic pond with floating plants: hyacinth ecosystem receiving treated sewage and building up bottom muck.

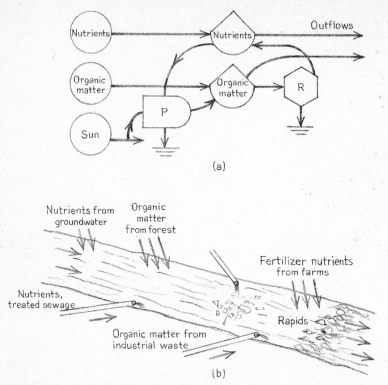

(a)

(b)

Figure 7-8 A steady state with inflows and outflows of nutrients and organic matter. This pattern is typical of most ecosystems. (*a*) Energy diagram. (*b*) Example: a typical stream.

eutrophic lake covered with rapidly growing floating plants, water hyacinths. The partly decomposed remains are carried from the hyacinth ecosystem and deposited on the bottom of the lake or carried downstream (yield).

Figure 7-8 is the most typical ecosystem, with inflows of both organic matter and mineral nutrients so that both photosynthesis and respiration are stimulated to higher levels than they would reach without the inflows. Either may be in excess at any particular time. Such streams are rich in many kinds of activity.

Many streams may receive organic matter from their watershed or from city waste pipes. With more organic matter flowing in, consumers have two sources of food—one from plant production using sunlight, and the other from the special inflow. The consumers thus put out more nutrients as by-products than they would if there were a regular, closed producer-consumer cycle. The excess nutrients flow out so that the system can exist in a steady state. Such a pattern—an extra inflow of matter and a net release of nutrients—was shown in Figure 7-6. Many estuaries are ecosystems of this type, receiving considerable organic matter from inflowing rivers and exporting minerals to the open sea with the outgoing tide.

Most ecosystems have some inflows of both nutrients and organic matter; consequently, most ecosystems are a combination of Figures 7-6 and 7-7, as in Figure 7-8. A marsh, for example, receives some nutrients, and some organic matter when waters overflow, but the outflowing water may carry out some of both as exports.

GROSS PRODUCTION, NET PRODUCTION, NET GROWTH, AND YIELD

In Chapter 3 we learned the principle of survival among competitors: those which survive are those which can apply their stored resources as feedback to pump in more energy or exchange them externally for more energy. Thus, most systems that survive have feedback pumping work pathways (see Figures 3-2, 3-3, 3-4, etc.). Figure 7-9 shows the standard pattern for a self-maintaining system with a feedback pathway, the usual depreciation pathway, and a pathway of export of energy from the main storage of order to another system. Some work is shown feeding back in exchange for the export. The diagram includes some commonly used terms, whose definitions follow.

Gross production is the flow of organic matter at G as a result of the main production process by which the energy from the source (E) is transformed to a higher-quality form. The production process has feedbacks (F) from storage (Q) and from outside the system (X) as obtained by exchange. The net energy

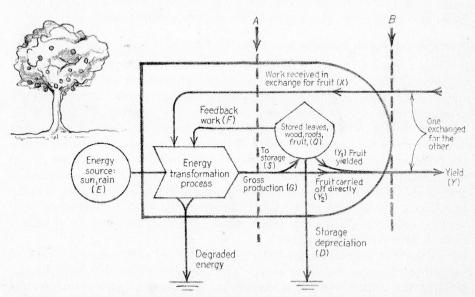

Figure 7-9 Definitions of *net* and *gross production* for a typical ecological system. Example: a tree that yields fruit and food to insects, which carry it away. The tree receives work in exchange for the yield of fruit. Several kinds of net production can be defined: net production of process unit (A, which is G minus F and X); net yield or net production of the whole tree (B, which is $Y - X$); and net storage contribution (Q, growth).

moving from the left across the boundary (A) is the difference between the flow to the right (G) and the return flows (F and X). This we call *net production of the process at point A.*

Yield (Y) is the export of high-quality energy products from the system. This yield is in two parts: (Y_1) is the yield that comes from storage; (Y_2) is the gross production exported directly as yield without first being stored. The *net yield* is the *net production of the system* calculated at B: it is the difference between all the yield (Y) and the high-quality work received back in exchange (X) across boundary B.

Another kind of net energy is the *net storage* (*net growth*) of high-quality energy (Q), which is the difference between the rate flowing into the tank (S) and the three outflows: feedback work (F), depreciation of the storage (D), and yield (Y_1).

For example, Figure 7-9 shows a farmer who has some apple trees. The trees make food (leaves, roots, fruit, etc.) using the energy of sunlight. The production output of this photosynthetic process is gross production. However, the tree is using some of this new food as fast as it is made, because the tree has to feed its own cells, leaves, tissues, trunks, roots, etc. Thus, by the end of a twenty-four-hour period, much of the food made by the trees is used up by them as the energy source for their own maintenance, work, and feedback of energy efforts to catch and process sunlight. If a tree makes more food than it is using, it will grow. This is the net production of the process. Suppose that the farmer is picking some fruit off the trees regularly as a harvest yield. This yield is net production of the system of apple trees.

Producer-Consumer Systems with Yield

In Chapter 1 we discussed a farm which exported its food products as a yield and, in order that it might do so, was supplied with fertilizer and other materials and services. Many ecological systems without man are also exporters of yield to other ecosystems or to man. For example, the Silver Springs River in Florida develops rich growths of plants and animals, some of which swim or drift downstream to feed organisms there. Many cold wetlands, for another example, build up dead organic matter, so that the living vegetation must grow on top of it, gradually building its base higher and higher on top of accumulations of peat. In effect the ecosystem is exporting organic matter from its system to the zones it is leaving behind below. Ultimately, the peat may become coal. Another example is a sea that sends salmon migrating upstream as an export from its productive food chains.

Often man becomes a harvester of the yields of ecosystems, as when he cuts timber, harvests fishes, or cuts peat for fuel. (See Figure 4-5, for example.) If yield was taken without a compensating inflow of new nutrients, the necessary nutrient elements would be drained out with the yield and the system would become impoverished and eventually nonyielding. Early farming by American pioneers depleted the soils so that the farmers had to move further west. The lands that were abandoned collected their nutrients again from rain and from weathering of parent rock. In modern agriculture, the nutrients are supplied as fertilizer. In systems that are not under human management, yields develop when there are regular nutrient supplies—not when there are no special inflows. Streams can develop yields of fish because the streams receive nutrients from draining their watersheds.

SUCCESSION OF SPECIES

The main trends in storage of organic matter and mineral nutrient storage, in production, and in consumption are facilitated by organisms—plants, animals, and microbes. Because of widespread seeding of plants, spreading of microbes and egg cases of microscopic animals, the organisms that are adapted to various energy conditions readily find access to ecosystems. Their means of starting a population are in soil, water, and air. Trends in energy conditions bring out successive surges of growths of different species, since different species compete best at different stages of transition. Furthermore, species are adapted to germinate, hatch, and start their life cycles when conditions for their growth are good. The sequence of living populations that appears in a transition to a steady state is called *succession*. The last populations in succession are those that replace themselves and become part of the steady-state condition. There are successive substitutions of species as specialists of different kinds appear at different stages in the succession. Large animals migrate in when the habitat becomes suitable.

An example of succession in an ecosystem that starts with an excess of organic matter is the sequence of life in a broth of boiled hay put out in an open bowl in a window in a schoolroom. First come populations of bacteria; next some small protozoa that eat the bacteria; later come large protozoa that eat the smaller ones. As organic matter is consumed and nutrients are generated, microscopic green plants ultimately take over in a steady state based on light energy.

An example of succession in an ecosystem that starts with an excess of nutrients is a bowl of fertilizer water seeded with a pinch of soil and some lake water and placed in a window. First comes an intense growth of very green algae. Later come a variety of other algae, small consumer animals, and the greater variety of mixed life that one finds in a pond. Early species are specialists at fast growth and net gain. Later species are specialized to do effective work of maintenance and control without much net growth beyond replacing their own structures. Species collectively maintain soils, establish nutrient storages, set up pathways of processes, store information for inheritance, etc. In general, where succession starts with a low initial state there is a period of mass growth with low diversity, followed by diversification and greater variety.

TRANSITION AND STEADY STATE

In Chapter 5 we showed how energy determines the shape of graphs of growth. In the water tank shown in Figure 5-2 the shape of the graph of water level depended first on the initial storage and then on the steady-state level for that energy supply. The general P-R model given in this chapter has similar graphs that trace the trend from the initial storage to the steady state. (See Figure 7-10.) When the initial storage has high concentrations of nutrients and small quantities of organic matter (Figure 7-10*a*), mineral nutrients are diminished as they become bound up in the organic matter that grows. When, on the other hand, the initial storage is high in organic matter and low in mineral nutrients (Figure 7-10*b*), the organic consumption process goes fastest, reducing the organic storage and releasing mineral nutrients. If the amount of matter in mineral form and bound up in organic matter is the same in the two starting situations, the steady states will be the same, as is shown in Figure 7-10*a* and *b*.

In Figure 7-10*c*, an external supply of nutrients is added which allows the system to develop a higher level of organic matter before it is limited by the amount of nutrients. In a bare field nutrients are gradually caught from rain

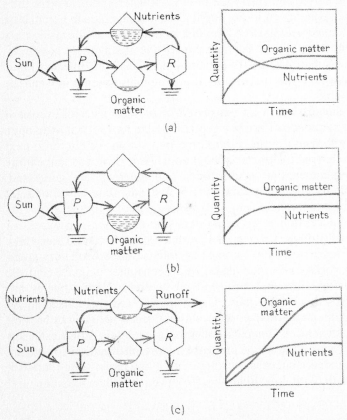

Figure 7-10 Growth curves for producer-consumer ecosystems for three different initial conditions. (*a*) Initial state high in nutrients. Example: an aquarium started with fertilizer. (*b*) Initial state high in organic matter. Example: a terrarium started with leaf litter. (*c*) Initial state with little storage; outside source of nutrients. Example: fertilized fishpond.

and from erosion of rocks to support the organic matter of an ecological system ultimately developing into a forest.

The steady state that develops in ecosystems is sometimes called a *climax*. It is the pattern of storage and processes that repeats itself from year to year. Many situations are so frequently interrupted by changing conditions that they never have time to develop a steady state. Trends toward one steady state are often interrupted with new inflows that cause the system to move toward different steady states.

EUTROPHY AND OLIGOTROPHY

The word *eutrophy* has become a part of our everyday language for discussing pollution of natural waters, although it is often not used as it was first defined for scientific usage. A *eutrophic lake* is defined as a fertile lake, meaning one with high rates of photosynthetic production and consequently also a high rate of consumer activity. It is an ecological system with a high rate of energy flow. Much of the world's waters are naturally eutrophic without receiving human wastes. Such lakes are green with algae or bottom plants but have adapted harmonius, successful, and continuing patterns of animals and plants.

The contrasting term is *oligotrophic*, meaning "infertile." Oligotrophic lakes are those with little nutrients and thus, often, with clear waters that are suitable for water supplies but do not have much living activity. Many efforts were made to fertilize oligotrophic ponds and lakes in order to get more yields of fish. In recent years there has been so much fertilization of lakes by waste that oligotrophic waters are becoming scarce.

Both eutrophic and oligotrophic waters have their place. Eutrophic lakes grow more fish, but not of the type that appeal to sportsmen. Oligotrophic lakes have a smaller total production of fish, but those that do develop tend to be the type that move vigorously around over a wide area seeking food and readily taking hooks.

Some popular writing describes eutrophic lakes caused by wastes as being dead. This is incorrect: a eutrophic lake is one that is intensely alive. The lake may be so alive that spurts in growth develop too many consumers; then, during periods of cloudy weather, respiration may exceed production to the extent that the oxygen in the water is all consumed and there are fish kills. Newly eutrophic lakes, unlike older ones, do not have animals and plants adapted to survive oxygen variations. Also, lakes made eutrophic by wastes may contain toxic matters, which sometimes interfere with the development of a balanced ecosystem.

PRODUCERS AND CONSUMERS FORM ZONES AND VERTICAL PATTERNS

Since patterns of growth tend to be located near their energy sources, producers and consumers tend to be arranged in ecosystems according to the

"Never mind the weather report. What's the eutrophication report?"
Sidney Harris, The American Scientist, September-October 1971.

locations of the inflow of sunlight and the development of organic matter. In lakes, the producers—algae and other plants—tend to be at the top within lighted zones. The consumers—microbes and animals—tend to be underneath in deeper water or at the mud bottom. Forests are similar: the producers are the leafy crowns up in the sunlight; the consumers are the roots, trunks, animals, and soil organisms underneath.

Gravity is used in these systems to transport products down to the consumers, but other mechanisms are required to return the nutrients back up to the producers. Evaporation, by the sun, pumps water up through the wood tubes in trees. In aquatic systems, there may be turbulent circulation due to winds, currents, or waves. In other systems there is a movement of some animals that migrate up and down, eating in one zone and releasing wastes at another: for example, fishes move up at night and down in daytime in lakes and in the sea.

DIVERSITY

In Chapter 6, and in Figure 6-10, we considered that energies adequate to develop order go initially into establishing a system. Next, they go into diversification to increase the effectiveness of use of energy and to tap auxiliary, lesser sources of energy. In ecological systems, diversity takes the form of many species, with their specializations for adaptation. A forest, for example, has a different insect for each taste of leaf.

We can measure diversity by counting the number of *types* per standard sample counted. For example, one may count twenty species of trees in a sample of 1,000 trees. For trees, diversity is considered low if less than five types are found in 1,000 units counted. Diversity is high if more than thirty types are found. Diversity is often used in considering whether a system is using many sources of energy or only a few. Diverse systems have a division of labor and specialization, so that they may use their energy sources more efficiently. For example, a diversity of plants in a forest helps to use all the sun, nutrients, and water. After succession, the amount of diversity ultimately developed depends on the energy available to support diversity after other priorities are met.

In climates that have sharply differentiated seasons, like the alternation of summer and winter in high latitudes and the alternation of wet and dry seasons in the tropics, those ecosystems are best adapted which have ways to fit processes of energy use to the energy seasons. For example, many trees drop leaves during cold or dry seasons, and put them back when warmth or rain returns. Many animals go into inactive periods when the habitat is unfavorable. In the sea, microscopic plants (algal plankton) shift species with seasonal changes in the supply of nutrients. All these adaptations to changing seasons require energy for special structures, biological processes, and the substitution of species. Changes in seasons cause variety to be large over the seasons instead of at any one time. In some tropical areas, where seasonal changes are less, diversity tends to exist all the time, with many adaptations to organize species so that they work well together. Usually, species adapted to a high-diversity situation are those with elaborate behavioral means (such as bright colors and special sound signals) for isolating their own roles without being confused by others.

Because some systems start with large storages of energy and raw

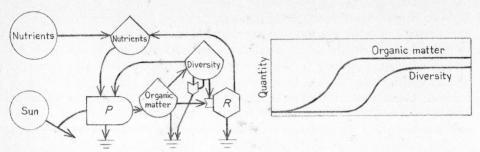

Figure 7-11 A pattern of development of diversity and change during growth and succession where nutrients are flowing in.

materials, their total development in the successional period may be higher than the steady state can support. (We have already considered this in Figure 5-8.) Then there may be more diversity at the crest of the growth than later.

The diversity of species can be limited by the number of different species available in an area. In estuaries, the tide usually supplies an enormous variety of organisms as seeding; but some environments, such as isolated islands, may have relatively little access to populations of plants and animals. The ecosystem of an isolated island may therefore have less variety than that of a less isolated island. Since the transportation of new kinds of plants and animals requires energy, isolated islands have less variety because there is inadequate energy for transportation. Species that become extinct are not immediately replaced.

The pattern of increasing diversity in one kind of succession is shown in Figure 7-11. The diversity of species is one form of storage of high-quality energy; it is a form of ordered information which costs energy to maintain but has an energy-helping role in feedback. Diversity increases uses of energy by feedback because it takes several stages of conversion of energy to generate variety.

Self-maintaining unit with two energy sources

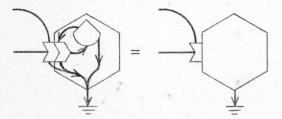

In Figure 7-11 we use the consumer hexagon symbol in a new way. Where more than one type of energy interacts as input to an energy-consuming unit, it is sometimes useful to show the interaction symbol coming out of the consumer hexagon, without showing the interior detail.

Stress and Diversity

If there are sudden actions that deplete the energy storages or inflowing resources of an ecosystem, the system is deprived of energy that would otherwise have been available to do more work and build more structure and diversity. Thus, such stresses as harvesting and exposing a system to toxic chemicals tend to depress the general level of structure. Often, this shows up as a decrease in the diversity of species.

If the stress effect continues, there may be a substitution of organisms that are adapted to live with the stress or even to get some energy from it. For example, a strong current that washes one species away will be used by a better-adapted organism to get more food. In adapting to special energy effects that would be stresses if there were not special adaptations, diversity of body functions may replace diversity of species. Many ecological systems in high-energy situations have a low level of diversity of species, but a high level of diversity of special adaptations within species. For example, there are only a few species on a beach, but those there are have many special adaptive aspects requiring energy. Coquina surf clams and sand crabs are examples: they can shift with the tide and continually reburrow into the sand.

SUMMARY

In this chapter we have studied how flows of energy develop order in ecological systems in the general symbiotic pattern of plant producers and dependent consumers. The production-respiration (P-R) pattern was examined and diagramed to show the kinds of interactions involved in the mineral cycle, the resulting growth curves, and conditions of high and low flows of energy due to high and low levels of nutrients that are commonly called *eutrophy* and *oligotrophy*. Variations in the basic producer-consumer pattern include the system with nutrient supply and organic yield and the contrasting system with organic inflow and nutrient export. Consumers were diagramed simply as one overall process, or divided into plant consumption and other consumers, or further divided to indicate the food web of interacting consumers and several stages of transfer of energy. Finally, some of the more detailed aspects of ecosystems were considered, such as zonation, diversity, and changes during growth and succession. At this stage, the reader should have some general concepts about the common characteristics of ecological systems and the manner by which the flow of energy in food chains and mineral cycles organizes and stabilizes the patterns of nature that cover much of the earth. The biosphere as a whole has the same overall characteristic symbiosis of producers and consumers, and these, working smoothly together, form the basis for life. As part of this, human beings have developed the roles of service to the biosphere on which they are still dependent.

Chapter 8

Energy Flows
of the Earth

The principles by which energy flows and generates order apply to all scales of magnitude in the universe, from the world of tiny molecules to the vast systems of stars. Man's basis on the planet Earth is the flow of energy that operates the weather of the atmosphere, the currents of the oceans, and the great earth cycle that builds and erodes mountains. These cycles accumulate resources and maintain them in the ground—coal, oil, iron, fertilizer chemicals, and all the other material and energy bases for our civilization. To understand how human activity is limited, we must understand the energy flows of these overall earth systems.

LIFE-SUPPORT SYSTEMS

Directly or indirectly, all the great systems of the earth run on solar energy. People who say we do not use the sun's energy are wrong: all the sunlight's ability to do work is being used to make winds, waters, and earth processes go. Sometimes we call these great cycles and processes our *life-support system*. When the space program sent men to the moon and back, the most important and costly parts of the equipment—and the ones that ultimately limited the time the men could spend in space—were the many items of machinery and stored goods necessary to provide a life-support system for the astronauts. Food and oxygen were required, and a system was needed to absorb waste products: carbon dioxide, urine, and feces. (See Figure 7-1.) Air and water had to be kept clean. We take our life-support system on earth for granted until we try to make even a small one to do the same job.

The great life-support system of the earth not only keeps the air, water, and land surfaces continually renewed, but so far it has had enough capacity to absorb much stress, poison, and wastes flowing out from our cities. Only locally have accumulations of harmful matter been serious. The life-support system generates soils, water cycles, and chemical processes that are necessary to produce food. The huge energy flows in these worldwide processes are never paid for with money. Our economic system recognizes with money payments only the energy pathways operated by human beings. Thus we often fail to realize how large and essential the life-support system of the biosphere is until we must try to do without it or until we begin to overload it with overdevelopment. Consider the air, oceans and land cycles: each of these subsystems of the biosphere has the characteristic patterns of all energy systems—stored high-quality energy, the feedback of stored energy into pumping actions on the main flows, and the recycling of materials.

SUNLIGHT

Light radiating out from the sun into space strikes the earth as was shown in Figure 2-2. This inflow is a mixture of rays of different intensities of energy. Some are invisible high-energy rays which give us sunburn; we call these *ultraviolet rays*. Of the rest, half are *visible* and intermediate in energy, and half are invisible rays of weak energy that we sometimes call *infrared rays* or *heat radiation*. Figure 8-1 shows the distribution of the incoming sunlight type of energy.

The ultraviolet energy is largely absorbed in chemical reactions with chemicals in the atmosphere and on the ground. These rays even sterilize some

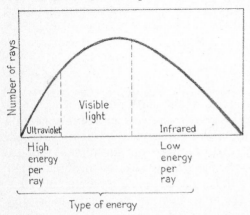

Figure 8-1   Solar energy reaching earth.

bacteria. The visible light is used by plants in the chemical reactions of food manufacturing that we called *photosynthesis* (see Figure 7-1). The infrared energy is mainly absorbed by the surfaces of the earth or seas, which become heated up in proportion to the heat radiation absorbed. Some rays of all types bounce off shiny objects like tin roofs and especially off the tops of clouds and go right out to space again. Of the sun's energy received on earth, 25 percent bounces back without doing work on earth.

Products of chemical reactions and photosynthesis are among the ways in which the sun contributes energy to the large-scale systems of the earth. Food produced by photosynthesis that is not used up again by microorganisms, animals, human beings, or other consumers ultimately is buried and stored to form coal or oil. Woody plants are buried as fibrous peat that eventually is pressed and heated and converted into coal. Soft-bodied aquatic plants called *algae*, when buried and exposed to heat and pressure, are converted into oil.

Calories of heat absorbed at the surface of land and water from heat radiation and from the other rays as by-products of their reactions have ability to do work. The amount of such Calories is large; but since we said earlier that dispersed heat energy is low-grade, we should explain here when it is or is not possible to get work out of heat. One can get work from heat only when the heat is unevenly distributed. Steam engines run on the same temperature-difference principle as the winds of the atmosphere; we sometimes say, therefore, that the atmosphere of the earth is a giant "heat engine."

Potential Energy between Places with Differences in Temperature

In order to understand the great systems of the biosphere we must understand the principle of the "heat engine." Remember that we have described evenly distributed heat as a low-grade energy which is not usable for doing more work. The molecular motions are so evenly distributed in so many directions that there are no ways to organize them to do larger-scale work without using additional energy sources. The term we use for heat concentration is *temperature:* if there are no differences in temperature, heat is evenly distributed. And when there are no differences in temperature in a source of energy, no large-scale work can be obtained from it.

However, if there is a difference in the concentration of heat (a difference in temperature), then that fraction of the heat which is different can be used to do work.[1] It may be helpful to examine Figure 8-2 on the supposition that the two columns of storage at *A* and *B* are water. The part of the two water storages that represents a difference in stored energy and hence in the ability to cause a flow is the difference in height of the two columns of water. A flow of heat and a flow of water are similar: both use a difference in stored energy.

In Figure 8-2, compare the situation where there is no temperature difference with the one where there is a difference. The ability to use heat energy depends on the temperature difference. Notice that the heat energy which can be used is the percentage obtained by dividing the temperature

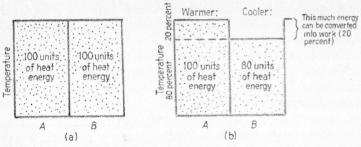

Figure 8-2 Fraction of heat energy that can be used to do work using the temperature difference between two places *A* and *B*. (*a*) With no temperature difference, no work can be done. (*b*) The percent of heat usable for work is the fraction due to the difference in temperature.

difference by the total temperature (see footnote 1): here, the result is 20 percent.

Wherever temperature differences develop owing to the heating of the atmosphere by the sun, flows develop using the potential energy. In this manner sunlight drives the ocean currents and the winds of the air. A process driven by the energy in a temperature difference is called a *heat engine*. Human beings build heat engines (steam engines, automobile engines, jet turbines, etc.), which also use the potential energy inherent in temperature differences (Figure 8-3). We will discuss this subject further in Chapter 10.

ENERGY SYSTEM OF THE ATMOSPHERE: WINDS

Wind is made by temperature differences in the air. This is true of a gentle evening breeze and a roaring hurricane. The main energy driving the flows of winds in the system of the earth's atmosphere, storms, and patterns of rainfall is the temperature difference between the tropics and the polar regions. Because the tropical regions of the earth are more perpendicular to the sun and are exposed to sunlight for longer hours in the course of the year, they get much warmer than the polar regions.

The heat energy that is part of the difference in temperature is the energy source for driving the general circulation of the air. This works as follows. In warm areas of the earth, the air is warmed so that it expands. This makes it rise and push up the top of the atmosphere (Figure 8-4*a*). The air above the tropics, being "taller," tends to flow over toward the poles. Some comes down at mid-latitudes, closing the circulation. The rest goes to the poles. In the polar area the air is cooled and tends to sink and contract into a heavy, dense, shallow mass until its weight is enough to cause it to flow from the polar regions toward the tropics again. Because of the way the earth rotates, the air circulation has additional spirals and bands, but basically it is the difference between tropical and polar temperatures which causes the winds and their motions.

As the cold polar air forms and moves southward, it comes into contact with warm, moist tropical air, making a sharp temperature contrast capable of

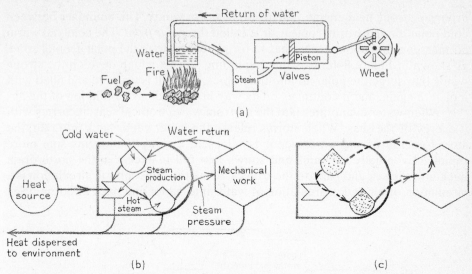

Figure 8-3 Energy flows in a steam engine that converts a temperature difference into mechanical work. (*a*) Steam engine. (*b*) Energy flows. (*c*) Circulation of water.

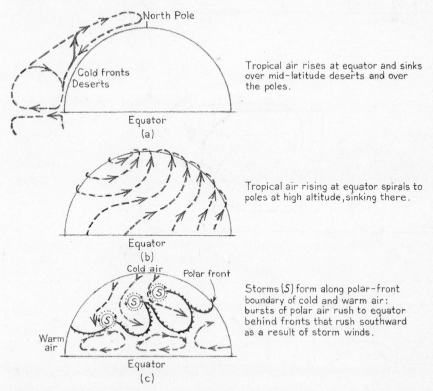

Tropical air rises at equator and sinks over mid-latitude deserts and over the poles.

Tropical air rising at equator spirals to poles at high altitude, sinking there.

Storms (*S*) form along polar-front boundary of cold and warm air: bursts of polar air rush to equator behind fronts that rush southward as a result of storm winds.

Figure 8-4 Main system of the atmosphere. (*a*) General circulation, side view. (*b*) Upper winds. (*c*) Winds at ground, showing storms that develop along the boundary of the polar front and send surges of cold air to the tropics.

driving a strong heat engine. The result is storm winds. The boundary between cold polar air and warm tropical air is called the *polar front*. The cold and warm air masses spiral about each other to form storms along the polar-frontal zone, as shown in Figure 8-4*c*. It is these spiraling storms which develop the strong winds that surge the polar air into the tropics. There it is warmed and filled with moisture again, and it continues on its way in the endless circulation of the air.

When water evaporates into the air from warm tropical seas, it carries with it energy of the seas. When storms release the water vapor back into rain, the stored energy of the water vapor is released, making the storms spin more rapidly. Eventually the rains and snows that fall in excess in the north drain back into the sea and back to the tropics. Thus the atmospheric circulation is a circulation of air and a circulation of water into vapor and back into rain. (See

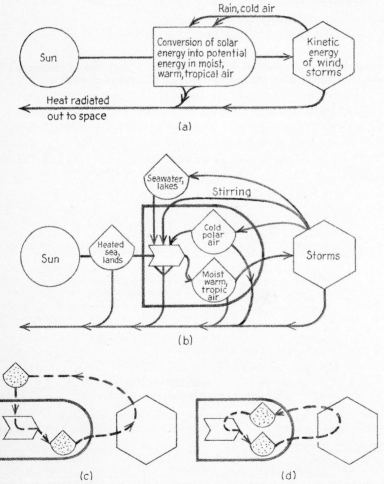

Figure 8-5 Main energy flows and material cycles in the atmosphere. (*a*) Simplified summary of energy flow. (*b*) Energy details. (*c*) Hydrologic cycle. (*d*) Air cycle.

Figure 8-5a.) We call the water-rain circulation the *hydrologic cycle.* (See also Figure 3-4.) Many kinds of special works depend on the hydrologic circulation, including manmade hydroelectric power plants and the geological work of rivers in carving the lands. Storms with high wind velocities are at the end of the chain of transformations of energy quality, and these storms feed back strong winds that make the worldwide circulation effective. Using the energy diagrams in Figure 8-5, we may simplify the atmospheric circulation as a single energy process. Notice that the pattern of this system is similar to that of others which have a chain of energy transformations that develops high-quality energy at the end and feeds it back to facilitate the flow. (Note the steam engine in Figure 8-3, the farm and town in Figure 4-1, and the examples of producer-consumer systems in Chapter 7.)

The atmospheric system uses small differences in temperature to generate large-scale general movements of air. The circulation of these large, gentle air flows causes effective conversion of energy into wind because it brings very moist tropical air into close contact with cold, dry air at mid-latitude. The energy of these two air flows concentrated close together causes circular storms of wind at the point of concentration. The storms in turn help to bring in more air from the tropics and the poles to interact further. The storms act like ball bearings in making the general circulation better. The whole weather system uses a large pattern of circulation together with small, intense storms to maximize the use of energy and thus to survive over alternatives.

As a life-support system for man, the winds are responsible for our climates, for recycling water to produce necessary rains, for dispersing concentrations of impure air, for turning windmills, for eroding the dry mountains of the deserts, for helping to make soil, and for driving the waves and currents of the oceans. Life as we know it would come to a stop without the weather system. People often talk of diverting this energy to some other use, forgetting that our existence is already dependent on its present use.

CURRENTS OF THE OCEAN

The oceans are a major part of the energy system that supports life. Stretching from poles to equator, the large storehouse of the planet's water moves in continuous circulation with a characteristic, similar pattern in each ocean. Figure 8-6a shows a simplified pattern of currents of the Atlantic Ocean, with water moving toward the pole on the left side and returning on the right to the tropics. Mainly, the flow of ocean currents is driven by the flow of winds above the water. The winds make the sea surface rough with waves and then press on the water surface until the sea is in motion in the same direction. The sea whirls in the same circle as the average wind.

The other source of energy which makes and moves ocean currents is differences in water temperatures. The seas in the tropics, receiving the tropical sunlight all year, become heated. This warmth is transported by the currents to the poles, where the heat is in sharp contrast to cold waters, cold air, and

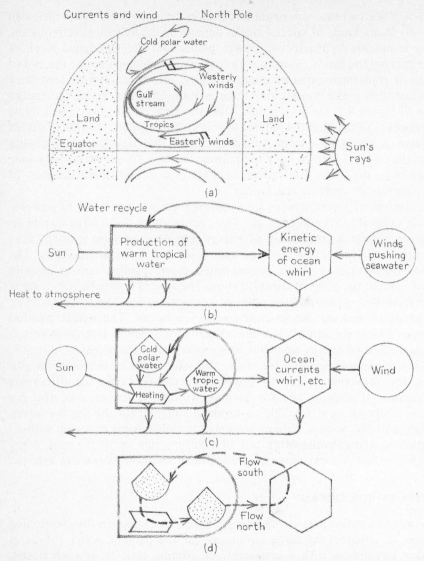

Figure 8-6 Ocean circulation, simplified. (*a*) Currents and wind in the Atlantic Ocean.
(*b*) Simplified summary. (*c*) Energy flows in main currents. (*d*) Circulation of seawater.

cold lands. The temperature contrast causes additional motions in the water
currents as cold, dense water slides under warm water. Notice the counterclockwise cold polar current.

Figure 8-6*b* diagrams the energy flows of the ocean with energy derived
from wind causing the sea to whirl in a great circle. The sun is shown as
generating warm tropical water which then interacts with cold polar water and
cools to become cold polar water. The wind drag and differential solar heating

drive the sea. Both contribute to the large-scale counterclockwise whirl which is shown as a storage of kinetic energy. Ultimately, the energy of motion is degraded by friction into heat. The energy of the currents comes from the push of the wind and from the interaction of water masses of different temperatures. Much heat from the tropics is transported to polar regions, where it is lost into the cold water and cold atmosphere. This affects winds, as we noted earlier. Figure 8-6b shows the energy relationships and Figure 8-6d water circulation.

To summarize: this is another example of a system with a circulation of materials driven by inflows of potential energy which develop energy storages and feed back energy to facilitate the inflow of energy and thus ensure the survival of the pattern.

EARTH CYCLES

The greatest systems of energy of our biosphere are the flows of the earth itself in cycles of building and erosion of mountains. The order and patterns of the continents and the sea-floor basins are continually being formed and replaced with earth materials going around and around. The basic pattern may be like that of the other systems, and the energy flow diagrams seem to be similar; but here we are less sure of what is happening because so much of it goes on out of sight, many miles below us. Even the sources of energy involved are partly obscure. Figure 8-7 gives a simplified view of the earth cycles that fits the facts known so far. How steady these cycles are is much discussed. Some people believe that the earth is in a fairly steady state when considered over millions of years; others believe that the patterns undergo continual major changes without any kind of repetition. We cannot answer the question of how steady the cycles are, but the flows shown in Figure 8-7 are known.

Contributing to the energy sources is the flow of water. Water rains on the continents and erodes materials, carrying them as suspended particles to deposit them as sediment in shallow estuaries and seas near the shore. The weight of the sediment causes the land to sink where it is deposited. As the sediments are washed from the continents, the land becomes lighter and springs up, like a mattress when weight is taken off it, but this takes thousands of years. With land at the coast being pressed downward and land in the middle of the continents rising as its surface is removed, the deep earth in between is sucked and squeezed from the coastal area to center of continent to replace that being lifted away. The land makes a complete circle, as is shown in Figure 8-7, going from mountains down the rivers to coastal deposits, from coastal deposits downward to deeper earth, from deep coastal areas back toward the center of the continent, and finally upward to replace eroded mountaintops. This circle is called the *sedimentary cycle*.

Erosion by rains, in conjunction with heating by the sun and the action of vegetation, makes this cycle go. We see in Figure 8-7 that the continents rise up above the rest of the rock. Like corks on water, continents float higher than the blocks of rocks forming the sea floor because they are made of lighter (less

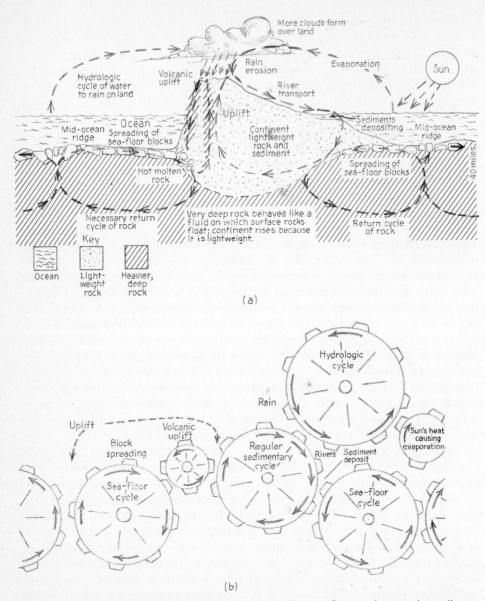

Figure 8-7 (a) Simplified view of earth cycles, including sea-floor cycles, regular sedimentary cycle that includes rock formation and mountain building, volcanic variation of mountain-building cycle, and hydrologic cycles that help drive the system. (b) Cogwheel analogy for earth cycles. Cycles of the wheels are coupled; one wheel turns with another.

dense) materials. The sedimentary cycle is mainly a cycle of light materials of continents going from light rock to sediment deposit and then reforming as light rock again. The effect of the hydrologic cycle is to move the earth in a circle from the land to the sea, then back under the land and up from below as new land.

As the lands are formed into sediments and then pressed and pushed back up under the continents over millions of years, the strata are pressed and loose sediments become rock. Heat contributes to the rock formation. The interior of the earth is hot: the temperature increases as we go downward, at about 38°F per mile. There are several sources of this heat, and these contribute to the energy available for the earth cycles. Some of the heat may remain from a time billions of years ago, when the earth may have been newly formed and molten. Some of the heat is released from atomic nuclear processes: there are tiny bits of such radioactive substances as uranium and thorium scattered all through the rocks of the earth, and the reaction of atomic decay frees heat, which accumulates. Relatively large amounts of radioactive elements become involved in the sedimentary cycle. Wherever there are movements that press rocks together, the pressure also increases the temperature. Remember that temperature is the concentration of molecular motions; therefore, pushing matter more tightly together increases the concentration of molecular motions and the substances involved become hotter.

Sediments that are buried and covered over contain organic matter and chemicals, like oxygen, that can react with the organic matter if enough heat is present to start the reaction. These substances contain potential chemical energy. It may remain stored in the rocks for millions of years before conditions are suitable for the reactions to take place. When pressure and heat are large, as in a zone where mountains are being pressed, the potential chemical energy may be released, generating heat and other actions.

As is shown in Figures 8-7 and 8-8, there are cycles of earth matter under

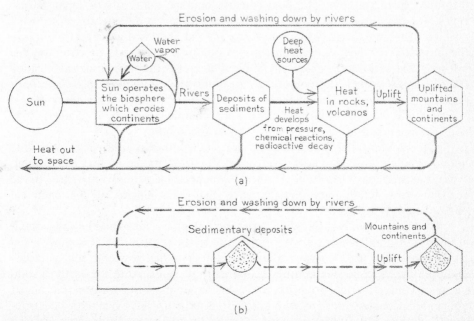

Figure 8-8 Energy and material flow of the earth. (*a*) Simplified summary. (*b*) Cycle of earth materials.

the sea that are coupled to the continental erosion cycles. Like cogwheels, they turn in opposite directions. The floors of the seas have blocks of rock that spread away from central ridge lines. Rock that emerges from below takes their place as the weight above is removed. The land that rises creates underwater ridges. On the edges of the oceans, the spreading blocks of the sea floor circulate downward when they reach the continental masses. Sometimes these plates go deep under the continents, and sometimes they hit with great pressures that generate molten conditions, volcanoes, and earthquakes. Apparently the pressures cause rock to return from the edge of the continents back to the center of the sea again, but no one is sure about this.

Wherever there are upward pressures, mountains form. Sometimes these mountains are made from volcanic squirting of molten rock, but most often they are made from the steady folding and pressing of sedimentary rocks.[2]

Figure 8-7 shows the main pattern of earth cycles and how they turn in opposite directions. Figure 8-8 summarizes the energy relationships, known and suspected, of the earth system. The energy of the sun operates the living cover of vegetation on land that catches the rain and erodes rock to form soil. Soil washes out to sea to deposit as sediments. Sediments build up and develop heat. Mountains and new land push up, completing the cycle.

Volcanoes

Because it is related to air pollution, we should examine the overall chemical reaction of volcanoes. Volcanoes are part of the mountain-building earth cycle; in Figure 8-7*a* they are shown as a step in the process of converting sediments into uplifted mountains. The heat causes chemical materials to separate into types. When molten rock, with its high-pressure gases and steam, reaches the surface of the earth, the volatile part (steam and chemical gases) goes out into the atmosphere. The molten part pours out over the lands or over the sea bottom. The solid part is rich in chemical elements that are sometimes called the *alkaline elements*; the volatiles that go into the atmosphere are sometimes called *acid elements*. Figure 8-9 shows the action of a volcano sending out acid volatiles and alkaline solids. Silica, a main chemical ingredient of most rock, soils, sediments, and beach sand, is part of the solids.

After cooling, the acid volatiles can react with the cold rocks. With the help of sunlight and vegetation, they erode the rock into soils. As a by-product, these reactions leave the main chemical elements of seawater free to flow to the sea in rivers. Thus they contribute to the maintenance of the chemical composition of seawater.

What is especially interesting about the volcanic release and its subsequent actions is the similarity of a natural volcano to the total action of our industrial civilization. Our urban power plants and other industries emit acid volatiles and, like the volcano, leave other elements to be returned to the earth as solid wastes. Thus, the air pollution of cities resembles the natural output of volcanoes. Concentrated volatile acids from industry have been very destructive to marble statues, to human lungs, and even vegetation. Volcanoes are also

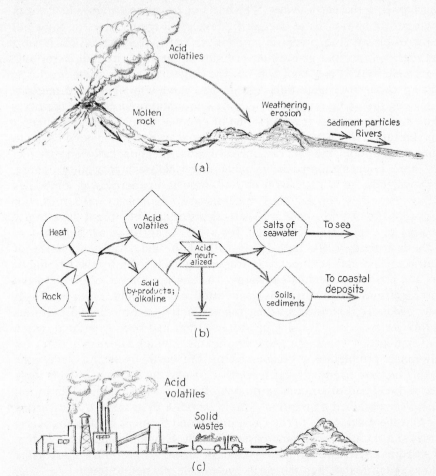

Figure 8-9 Chemical output of volcanos and cities has acid volatiles and solid by-products. Later, the acids are neutralized by reacting with the solids as part of atmospheric erosion. From the weathering, sea salts and sea sediments are produced. (*a*) Volcano. (*b*) Energy flows of volatile-acid reactions. (*c*) Human urban activity.

destructive to anything near them. However, the biosphere as a whole absorbs volatiles from volcanoes into its weathering process, and it can absorb some industrial output. Considered as a whole, human civilization is accelerating the earth cycles. Human activity may be helping to maximize the power flow for the earth.

Ore Deposits and Earth Reserves

The cycle of the earth develops deposits of iron, copper, uranium, coal, limestone, and all the other earth substances that are used in human economic activity. The flows of energy that turn the cycles cause special, rich deposits of many kinds to develop in various places.

Man taps into the earth cycles. The basis of the industrial revolution was the resources and reserves of the earth, developed from the great cycles of building, erosion, and movement of mountains. The concentrations of valuable material were part of the giant, but very slowly turning, circulations of soil, sediment, and rock. (See Figure 8-7.) The sun drives the water cycle by evaporating water that later falls as rain and erodes land. Eroded material deposits along the shore, and eventually new mountains form. Like the other cycles we have described, these systems of circulating rock and earth matter develop storages. Human beings discovered these accumulations and began to use them: the oil pools, the iron ore, the coal beds, the phosphate deposits, etc.

All these concentrations of valuable material are high-quality energy. Although we think of a petroleum or coal deposit as energy, we sometimes forget that the other concentrated minerals are also energy resources. For example, deposits of phosphate are an energy storage; we do not have to use other energy to concentrate phosphate for use as fertilizer in agriculture.

In the last century, the fantastic growth of the industrialized system, of agricultural yields, and of population was indirectly driven by the tapping of deposits and pools of accumulated energy. The energy storages of the giant and slow-turning geological processes were diverted away from the earth processes to a faster process of use and recycling, the new activities of industrial cultures. During this period man was using up storages that had been generated over a period of millions of years. He was using up the savings rather than living off the long-range production of resources by the earth cycles. As everyone knows, this fast use eventually must come to an end, as easily accessible large storages of fuels and other concentrations of material resources begin to run out. Humanity obtained a temporary, rapid increase in growth from diverting energy from the earth cycle to the cycle dominated by men; but our long-term future may not include further increases like that. Human beings are diverting the old mineral cycle through their system of cities. At present, we are using these flows much faster than the earth cycles can regenerate them.

How renewable are the earth resources? We are beginning to run out of rich concentrated earth resources close to the surface. In Chapters 1 and 4 we saw that the standard of living and inflation are controlled by the availability of rich resources. Basic fuels, coal and oil, phosphate for fertilizer, iron and copper for industry, and uranium are concentrated by the slowly turning earth cycle, which is driven in part by the action of the sun on the hydrologic and weathering processes, as was shown in Figure 8-8. New deposits are being concentrated and brought up from below by the earth cycle, but this process is slow. The rate at which we use oil has been faster than its rate of formation in buried sediments that are being compressed and heated. In a theoretical sense, the minerals are renewable, since they are being reconcentrated. But in practice we call these resources unrenewable because they are being reconcentrated too slowly to keep up with our present rate of usage.

In later chapters, we will see what kind of human life and human economy is possible when rich storages of accumulated minerals and fuels are mainly dispersed.

SUMMARY

In this chapter we have examined the energy flows in the biosphere and on and below the crust of the earth. Circulations of the atmosphere, of oceans, and of earth materials are driven in large part, directly or indirectly, by the sun. The sun's energy develops contrasts in temperature that causes heat-engine actions to develop. Energy diagrams of the earth systems show the same kind of energy chains we found in other systems. Production of energy first converts dilute solar energy which is then concentrated by consumer units in which the quality of the energy is increased and energy is fed back to facilitate the main energy flow. Mountains, storms, fast ocean currents, and carnivorous fishes in food chains are all examples of concentrated activities as the end of energy chains that return service by acting on the energy flows. Chemical cycles of the earth are turned by the earth's energy flows, producing deposits and resources upon which humanity is now mainly dependent. The chemical reactions of urban development have some similarity to those of volcanoes: both cities and volcanoes release acid volatile chemicals that are corrosive. The flows of energy of the earth make and maintain the patterns of the landscape and provide the primary basis for human life support which we often take for granted.

FOOTNOTES

1 This fraction, or percentage, is sometimes called the *Carnot fraction*, after the man (Nicolas Carnot, 1796–1832) who discovered this property of converting heat differences into work. The ability to use heat energy depends on the temperature difference, and the amount can be calculated as the temperature difference divided by the temperature of the hot place:

$$\frac{\text{Temperature difference}}{\text{Hot temperature}} = \frac{\text{Fraction of heat}}{\text{usable for work}}$$

To make this calculation, one must use the scientific scale of temperature that defines *zero* as the point of no heat energy (no molecular motion), rather than the common Fahrenheit scale that most people are used to. The temperature at which the molecular motion does not exist is sometimes called *absolute zero*. Absolute zero is believed to be minus 457 degrees Fahrenheit. The lowest temperature achieved in the laboratory is within a fraction of a degree of this. To calculate the Carnot fraction, add 457 to the Fahrenheit temperature used in the denominator.

2 Sedimentary rocks are those formed by the pressing together of sediment. The most common are shales, limestone, and sandstone. When temperatures are high enough, sedimentary rocks are changed into metamorphic rocks. Soft shale rocks may be pressed into slate; limestone rocks may be pressed into marble; and weakly cemented sandstone may be pressed into quartzite rocks. If the heat and pressure are great enough, all the rock melts and igneous rock results. The most common igneous rock that pours out over the surface of the earth from volcanoes is called *basalt* and is black. The most common igneous rock that emerges when mountains are built slowly under high pressures and temperatures is granite, a light-colored, very hard rock containing large mineral crystals including a great deal of quartz.

Part Two

Energy Systems Support Humanity

Looking back in history and across the vast expanse of the planet Earth, we must conclude that human beings are very versatile in adapting to an almost endless variety of energy flows and ecological systems. Because human beings developed social mechanisms for changing their cultures by responding effectively as groups, it was possible for civilizations to exist in various places using the various combinations of energy provided to man by the basic photosynthetic production of ecological systems of seas, lakes, grasslands, and forests. Part One discussed the fundamental principles of energy and the way energy flows develop and maintain order; Chapter 7 examined the general plan of production and consumption by which order is maintained in ecological systems. Early records of human life indicate that human beings were at first a minor part of these ecological systems; gradually they assumed a more and more important role. Human versatility was a means by which energy principles were expressed.

Human culture became a part—indeed, a controlling part—of the operation of the land, and of plant and mineral cycles. Human beings became partners with other life in the operation of the lands and waters. As time passed, human beings became more and more managers of their ecosystems, but they remained dependent on the ecosystems for life support. As human beings became more involved in energy flows, they became more and more capable of helping or hurting their own basis for existence—their partnership with the rest of nature. Since humanity and human culture are a component of nature and subject to the same laws of energy, our existence depends on the energy flows. In the five chapters of Part Two, we examine the energy basis for humanity in its environment.

Carrying capacity is a term used by people who manage lands for wildlife. The carrying capacity of an area is the population of animals—say, deer or quail—that the food chains of the ecosystem can support in a steady state. In a given area of land or water, the energy of the sun, rain, winds, etc., is fairly regular year after year, regularly supplying the energy basis of the food chains. When human beings were supported by food chains based on production by the lands and waters, human populations were also determined by energy in terms of the fertility of an area. We can speak of the carrying capacity of an area for human beings as well as for wildlife so long as forms of energy which support human life there are derived from the energy sources of the land. The kind of culture which exists also depends on the energy flows. If human beings utilize a large part of the food chains of an area, their populations can be larger: that is, the carrying capacity of the land for human beings is greater. If additional sources of energy are brought in from outside—as we import fossil fuels—the carrying capacity of an area is increased and continues to increase until new sources interfere with the sources already in use. For example, the use of solar energy is diminished by the paving of a parking lot. In Part Two we consider several human cultural patterns and the carrying capacity associated with them.

In our lifetime there has been a worldwide surge of rapid urban growth: this is the culmination of the industrial revolution that started in the nineteenth century. Much of the energy for the cities is now derived from coal and oil deposits in the ground. Because coal and oil were formed from remains of plants—ancient products of photosynthesis that were left unutilized by consumers at the time—we sometimes refer to these resources as *fossil fuels*. The use of fossil fuels is a relatively recent development. Before we study the energy basis for our life today, we should examine the energy basis for human life in earlier times, when fossil fuels were used only in a minor way.

Chapter 9 discusses early man in various cultures based on solar energy. Chapter 10 takes up the recent explosive growth of urban cultures on the basis of fossil-fuel energy. Chapter 11 examines the quantity and practicality of the energy sources which now support humanity or are proposed as sources for future support. Chapter 12 deals with our energy basis from an international point of view and suggests prospects for various countries as related to resources and technological development. Finally, in Chapter 13 we suggest how energy affects our individual life-styles and the faith we have in various alternatives. Part Two is, thus, historical: we examine the energy basis for humanity from primitive hunting and gathering societies to the surging urban societies, with their individual unrest, that are undergoing the energy crises of today.

Chapter 9

Energy Basis for Preindustrial Societies

Because the various climatic belts of the earth have seasonal differences in sunlight, rain, and access to migratory animals, there were many different patterns of existence among early tribes and early civilizations. In this chapter we examine some of the main bases of energy before the industrial revolution. Some preindustrial periods were stable for a long time; others were times of transition. Only for the last 10,000 years or so—the periods for which we have archeological and historical records—do we know considerable detail. The records show us the roots of our history and culture and suggest how solar energy, operating through the environment, can support humanity when there are no longer special, rich energy deposits such as fossil and nuclear fuels.

We can ask how much energy is necessary to support a human being; but a better question is: "What is the energy flow necessary to support the systems of which humanity is an integral part?" When a human population has become adapted to the available energy and environmental conditions, it develops characteristic ways of life, blending previous ways with new conditions. The word *culture* is often used to refer to the combination of a way of life, a language, various social interactions, government, religion, etc. The energy requirement for human beings includes the energy needed to maintain cultures.

ENERGY REQUIREMENTS FOR HUMAN BEINGS AND HUMAN CULTURES

The energy requirements for a human being include the food that he eats to run his body and the energy flows that support him with goods and services such as clothing and shelter. The energy for operating a human body on food is about 2,500 Calories per day in the form of carbohydrates and other foods that the body uses as fuel. However, to fully support a human being much more energy is required to provide shelter, clothing, absorption of wastes, social exchanges with other humans, flows of air and water, and so on. In the United States about 250,000 Calories of fossil fuel are used through machines for each person each day. To fully support a person, with all the inputs that are required for his culture, a whole life-support system is required. Very large amounts of energy are required to keep us going. A Calorie of modern packaged food from the supermarket has a fossil-fuel equivalent of about 10 Calories. Thus, food per person per day costs about 25,000 FFE Calories.

Many human needs are really those of the system into which the human being fits. As we learned in Chapter 3, if a system is to survive where there is competition, all its energy users must contribute back to help the energy processing of the whole. Variations in culture provide alternatives that compete for more general adoption on the basis of which are most effective for processing energy. Thus, a human being's survival depends on having a system into which his services can be fitted, in exchange for the life support he receives. All ideas and subcultures compete for a more dominant role in the human system. And humanity as a whole also competes with other combinations of species.

Figure 9-1 shows human needs from food and other inputs. Note that the supply of energy to a human being is in exchange for his high-quality services back to the life-support system; these services help make the system competitive. Because a person's role aids rather than merely drains the ecosystem around him, systems which include human beings can compete with systems which do not. If humanity misuses the sources of its life support, it will

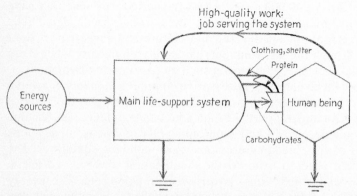

Figure 9-1 Energy requirements for a human being include the necessities—food, clothing, and shelter—and an opportunity to feed back work in exchange.

be replaced by other consumer life—perhaps algae, insects, and microorganisms that can tolerate harsh conditions.

Human Diet: A Carbohydrate Fuel and a Protein Source

Human dietary requirements are much discussed—from biology textbooks to television advertisements. Protein, minerals, and vitamins are required for good health, in addition to basic staple foods. Sometimes, in simplified discussions of world food problems, food needs are summarized as a carbohydrate energy supply and a protein supplement including vitamins and other necessary components. As is shown in Figure 9-1, good human nutrition can be shown as the interactive flow of both kinds of food energy. The carbohydrates have more calories but are of low quality. Proteins (such as meat), which take more energy to develop in the first place, are higher in quality but are needed in smaller quantity. Figure 9-1 also shows the interaction of food and other energy inputs in support of human activity.

Let us now consider several types of preindustrial human ecosystems. In these systems, foods, goods, and services were generated in a variety of ways.

HUNTING AND GATHERING SOCIETY IN A SEASONALLY STABLE TROPICAL FOREST

In a few places in the tropics rainfall, temperatures, and sunlight are so regular through the year and from year to year that a rich, stable evergreen tropical forest develops. At least until the present century, great forests stood without much change in the upper Amazon, in New Guinea, in the Congo, and elsewhere. These forests had a great variety of trees and other plants in lush profusion which supported tall forests. These supported a fantastic variety of animal life—thousands of species of small insects and other animals, though not many of each kind. This type of forest is called *rain forest;* very little of it now remains uncut. The rain forest is our best example of a complex, self-maintaining ecological system with much plant production, almost all of which is consumed immediately by a great variety of life. The trees, 300 to 1,000 years old, help to support many vertical layers of life. Light and rain are well used in photosynthesis; the products of photosynthesis are mainly used again by the system for its maintenance, as for replacing trees that fall. The minerals are cycled: they travel up the trees in water that goes up tubes in the wood to the leaves, and then travel back to the soil with falling leaves. After hundreds of years many forested areas developed a fairly steady state, growing, repairing, and continuing with little change, while turning over their materials rapidly in the process. Since there were never very large surges of net growth, and since there was little seasonal surge of production of fruits or animals, there were relatively small amounts of food suitable for human beings at any one place or time. However, there was a very regular and reasonably stable supply for the people who once lived in such forests.

In many places around the world, rain forests had tribes of people whose cultures were adapted to it. These people lived, in small groups, by hunting animals and gathering foods from the great variety of plants. Some remnants of these cultures have persisted to modern times and have been studied by anthropologists. Examples are the pigmies of the upper Congo River basin, some Indian tribes of the Amazon River, some tribes in New Guinea, and a primitive tribe of inhabitants in rain forests in the Philippines.

The energy basis of a hunting and gathering society in a stable rain forest is given in Figure 9-2. Such a society is a versatile feeder at the end of the food chain in the sense that it sometimes consumes larger animals. More regularly however, it derives foods from the lower stages of the food chain, including plant roots, smaller animals, and fishes.

Since human beings in deep, stable forests were taking only a relatively small part of the energy of the ecosystem, they put little load on the system and did not interfere with its stability. The population density was low: about one person per square mile. The carrying capacity of the area for human beings in this role was small; human populations were adapted by their birth practices and ability to survive disease to remain at a low density. By its own complex organization, the forest provided human beings a steady and stable basis for survival so long as their populations were low.

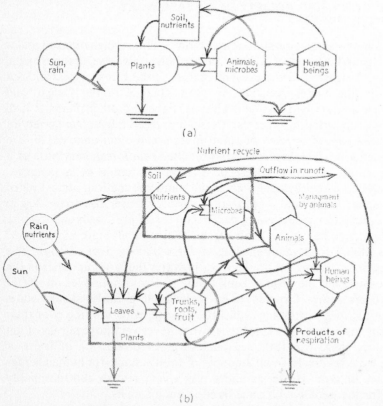

Figure 9-2   Energy basis for hunting and gathering society. (*a*) Simplified sketch. (*b*) Detail.

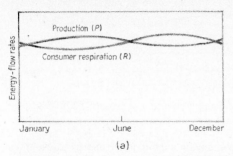 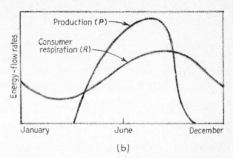

Figure 9-3 Comparison of seasonal trends in production and consumption for (a) regular climate and (b) sharply varying climate (polar region; winter dark).

In Figure 9-2, notice that maintenance and recycling in the forest are done by organisms other than man. In the sense of Calorie flow through man, the human role is small. Because people are spread out in the forest, it is no problem to absorb their wastes. Most human energy in this system goes into maintaining food supplies and shelter.

In other climatic belts, human populations expanded and declined with shifting zones of rainfall. However, in a stable rain forest, the human population remained fairly constant because the energy supply was nearly constant. See the fairly steady line in the graph of production in Figure 9-3a.

In earlier chapters we learned that most of the consumers of an ecosystem return services back to the system in exchange for food received. For example, animals recycle minerals, manage plants by their patterns of planting and harvesting, and help distribute seeds. Hunting and gathering societies in the forest may have had a similar high-quality role: helping to distribute seeds, controlling the populations of larger animals when they became overcrowded, and adding to the variety of situations for plants by developing clearings. See the pathways in Figure 9-2 which go from man back to plants and animals.

The human role is at the high-quality end of the energy chain. When the population is small, the total Calories of energy flowing through human bodies is small. However, even in hunting and gathering societies the quality of human work is high, involving individual intelligence, knowledge of the forest, and culturally transmitted information. In Chapter 6 we used energy cost to estimate the quality of energy flows. The energy used in the forest ecosystem to provide the base for human life is large. The feedback of human work to serve and control the system may also be large. By such actions as hunting and running off large and small animals, and eating nuts and fruits, humans controlled the kinds of species in the forest. In one sense the energy per person flowing in support of human beings and partly under their control was as great as it is in urban society.

The feedback of high-quality human work interacting with the low-quality energy of the sun develops the maximum possible work, according to the principle of interaction between high-quality and low-quality energy given in Chapter 6.

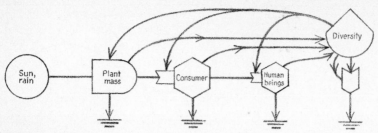

Figure 9-4 Energy basis and contribution of diversity. In exchange for energy to maintain diversity, there are feedbacks of work that increase the effectiveness and stability of energy flow.

Diversity, Stability, and Energy Costs

In Chapters 6 and 7 we used the term *diversity* to describe the variety of plants and animals that make up an ecosystem or the variety of occupations that make up a human culture. Where hunting and gathering society was being supported by a rain forest of great diversity, its food chains were based on many different kinds of fruits, roots, and animals. The diversity of foods required a considerable knowledge of the plants and animals of the life-support system. Since there was no single main source of food, the food sources were more secure against interruption by disease affecting any one plant or animal. The basis of human life was secure because the forest was diversified in its component plants and animals. If the climate is regular, the ecological system can put its energies into developing diversity of plants and animals. Such diversity makes the food chains to the larger animals more complex but also more secure against sudden fluctuations.

The diagrams in Chapters 6 and 7 represented diversity of plants and animals in one of two ways. One way is to show a separate symbol for each species and connect the symbols in a web of food pathways and other interactions; another is to show one storage-tank symbol for diversity, a quantity stored in the system (both ways are shown in Figure 6-12). Maintaining diversity requires much energy, since the various special properties that allow division of labor (these include special organs, chemical abilities, behaviors, and other inherited features of organisms) require considerable food energy to operate, maintain, and pass on to future generations. Diversity also requires more information, and that requires more energy. See Figure 9-4. However, diversity (both of species and of human work) allows specialization, and this contributes to the effectiveness of energy processing: note the feedback pathways in Figure 9-4.

HUNTING AND GATHERING SOCIETY WHERE GROWTH SEASONS ARE SHARPLY DIFFERENTIATED

In the far northern regions, where winters are long and harsh and sunlight is strong only in summer, ecosystems adapted by storing more organic matter. Energy was stored in animals, fruits, peat, or trunks of plants or sent southward with migrating birds and mammals. Then the system of life could resume each spring with the return of warmth, sun, and migrating animals. Diversity was spread over the seasons, with one species succeeding another in activity like players on a stage, so that more energy was seasonally connected and less was available for supporting diversity at any one time. In northern rivers, for example, there is the following sequence: explosive growth of algae and microbes, followed by aquatic insects and salmon, whose young pass through to the sea during one season and then later return to lay eggs.

Hunting and gathering societies adapted to the sharply seasonal pattern of nature by developing their own sharply seasonal patterns. They lived off the larger animals and fishes that stayed over the winter, or they migrated to the south, to the seas, from mountains to the valleys, or to the places where winters were milder, returning again to the places of seasonal productivity in summer.

Examples of cultures which adapted to northern climates are the Alaskan Indians and their European counterpart, the Laplanders—these adapted to the movements of salmon and caribou. Another culture, the Eskimos, adapted to the cycle of freezing and melting in Arctic seas and shores by using the large storages of overwintering seals, under-ice fishes, and bears. Houses were built out of ice (igloos) or out of peat banks shaped like teepees (Laplander houses). Further south, the Plains Indians followed the buffalo as their basis of life.

Production was for part of the year, but the consumers, including human beings, had to operate for the whole year. Figure 9-3 contrasts the patterns of production and consumption to which man had to adapt in the sharply seasonal climates and in the little-varying ecosystems.

Sharp seasons are also found in the warmer zones of the earth where there are sharply varying periods of rainfall. In such areas, foods vary with the seasons. The Australian aborigines are an example of adaptation to sharply varying dry and wet periods in warm climates. They moved long distances, adapting their hunting and gathering life to the semideserts where production was large during rainy times and small during the long dry periods.

SOCIETIES SUPPORTED ON SOLAR AGRICULTURE

Hunting and gathering societies had a relatively small population in a controlling role in the ecological system. With agriculture, larger human populations could be supported. In areas where human beings could substitute themselves and their farm animals for some other consumers, the wild system of food chains and diverse consumers was displaced. As was shown in Chapter 6 (see Figure 6-9), forms of energy developed from one source (in this case, hunting and gathering) can be used to develop a second source of energy (here, agriculture). Since that development was successful, agriculture, originally the second source, became the main source; hunting and gathering became the minor source. Occupying a greater part of the food chain allowed human populations to grow as dense as one to the acre (640 per square mile). The cultures of India, Egypt, and Panama in precolonial times are examples of dense patterns of agriculture.

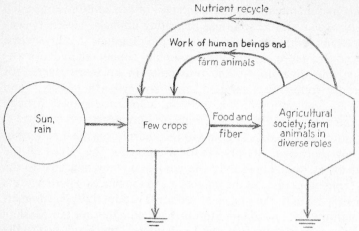

Figure 9-5 Humanity as main consumer, plant cultivator, and agent for recycling nutrients in solar-based agriculture.

Figure 9-5 shows the greater role of agricultural society as a main consumer with higher populations and more control of production. Such societies replaced the older pattern of diversity within a natural ecosystem. The diverse work of human beings on small farms was substituted for the diversity of species that had been there before: compare Figure 9-5 with Figure 9-2.

Where the climate causes regular surges of growing, the natural seasonal interruptions serve to cut back weeds and insects. In the northern climates, winter interrupts growth, giving human beings opportunity to use their own work early in the season and substitute their own crops for general vegetation. Similarly, in those tropical areas where there is a wet season and a dry season, as in the monsoon climates of Asia, human beings can use their work and irrigation to start crops ahead of the normal rush of growth of general plants during the wet season. Especially effective for solar agriculture are the delta areas of large rivers which flood once a year, providing good conditions for human agriculture. River floods bring nutrients, water, and new soils for a new crop. Dry periods in between the wet periods help to control wild vegetation, which otherwise would create too much competition. These changes also control insects.

Shifting Agriculture

Competing with wild ecosystems was not always effective. In rain forests, agricultural societies could compete with the lush natural growth only by cutting the forest and burning the logs and brush. Only two or three years of crops could be obtained from an area before plant growth and insects overwhelmed hand labor. Then, the tribe would move to another area and cut another plot so as to get another short period of growth. Such *shifting agriculture* is still found in many areas of the tropics, where the fuels and special products of urban culture are not available to control weeds and divert insects.

Shifting agriculture is shown in Figure 9-6. Here a human settlement moves from one area to another, taking energy in turn from each area. A village can be moved back to the first area after it has had enough years of regrowth. During the time of regrowth, an area captures nutrients from rain and rocks, builds soil structure, and develops shade plants that displace weeds and support different kinds of insects from those that would be most voracious eaters of the crops.

Shifting agriculture can be compared to a gasoline motor with several

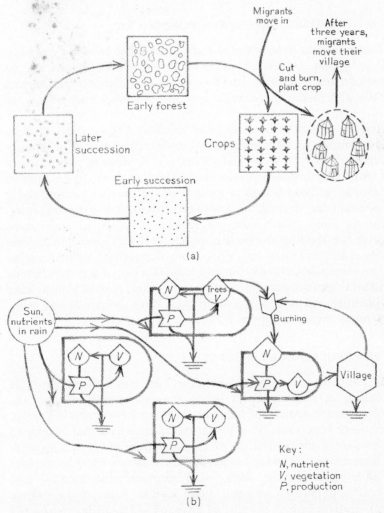

Figure 9-6 Energy basis for shifting agriculture in a tropical forest. The land goes through a cycle of use: people use it for a few years and then move away. (*a*) A map of plots in rotation. (*b*) Energy diagram.

cylinders. In each cylinder, explosions are followed by a quiet part of the combustion cycle as the cylinder regains its storages. The motor as a whole continues to run fairly smoothly because it shifts its demand for power from one cylinder to the next as each cylinder has its energy drawn off. The human tribal village, then, is like the motor, and its agricultural plots are like the cylinders. The village exists by shifting from one plot to another plot for energy, while the previously used plots are renewing their storages. Although each plot is undergoing succession and is forced to start again after the cutting and planting and harvesting, the overall landscape can be regarded as a steady pattern if the proportion of plots in the various stages remains the same.

SUPPORT FROM SEASHORES, LAKES, AND PONDS

Human societies have often been oriented to the energy bases of waters, seas, lakes, streams, and ponds. On islands, along estuaries, in lake regions, and on great rivers, preindustrial societies could live on food chains from aquatic ecosystems while also using the energies of tides, winds, and waves for transportation.

Many American Indian tribes lived on shellfish. Aquatic ecosystems provided a steady output, as clam beds and oyster reefs, available in summer and winter. Catching fish with weirs and spears was an ancient part of many tribal patterns. Aquatic ecosystems, like ecosystems on land, often achieve their greatest storages of food matter at the end of the growing season; these storages are the means for starting the spring growths with new reproduction. Fishponds were an even more intensive means of support. They were especially effective in providing protein. Fishponds were fertilized with organic matters from the land and with nutrients from human and animal waste. Fingerlings either were stocked from outside natural areas of reproduction or derived from natural reproduction in pond waters. How human beings were provided with shellfish and with fishes for protein from fishponds and other aquatic ecosystems is shown in Figure 9-7.

REGULATION OF POPULATION BY DISEASE

In ancient times, populations expanded whenever there was a rise in favorable resources, increasing carrying capacities. But there was apparently nothing like the current long-sustained growth of humanity that we have recently known. Population levels were kept much lower by various mechanisms that developed over the years as part of the adaptation of human beings and their surroundings. In many long-continuing cultures, population was regulated by various social systems affecting conception, birth, infant mortality, and the pattern of chronic disease. The patterns developed some kind of balance. So long as populations were relatively small, the role of diseases and parasites was one of testing the vitality of new infants and the aged, eliminating those whose

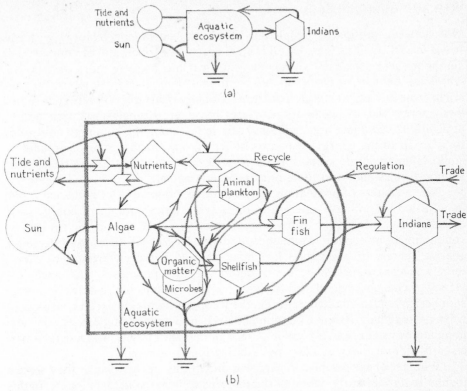

Figure 9-7 Energy flows to human beings supported by aquatic ecosystems. The term *regulation* indicates that size of population is controlled. (*a*) Summary. (*b*) Detail.

resistance was inferior or failing. Diseases tested the energy reserves and operations of the human body. They were an automatic mechanism for controlling the quality of energy and the size of human populations.

Epidemic disease was not a factor while populations were small and dispersed. However, when populations became crowded and did not have enough food resources, epidemics did develop. People who do not have enough to eat are more susceptible to disease. Epidemics greatly reduced populations, either restoring them to safer carrying capacities or causing people to move on so that they were better distributed in relation to energy resources. Epidemics were in a sense responses to malnutrition and to high population densities. Diseases were spread rapidly through such mechanisms as personal contacts and the infection of water supplies by human wastes, both of which become much more frequent when the number of people increases. Disease naturally seems bad to affected individuals, but diseases did have a role in adjusting populations to resources over the long range. Disease helped ensure the survival of the group by keeping the population adjusted to the supply of energy.

WAR AND BOUNDARIES

War between groups of people has characterized human history. War was found in early civilizations, is still found in primitive tribes, and of course exists among modern countries. In recent years war has become more destructive, with huge amounts of energy in "conventional" and nuclear weapons. We now regard war as bad, wasteful, and tragic for individuals directly affected. When less energy was available for warfare, however, struggles between human populations had some organizational role in adjusting the land area controlled by a group to the energy resources of that group. Each center of population could influence more distant areas only to the extent that its energies could support armies and send them out to spread control.

When centers of population were scattered, boundaries of influence would be adjusted according to energy. In times when large amounts of energy were available and effectively utilized, a population could spread its control by conquering or by economic influence to make a larger unit. In times of declining energy, centers of population lost influence and withdrew from each other; there was less conflict. War, if considered as energy processes, was a means of testing or changing boundaries of influence according to the realities of energy supplies. War also affected cultures, the training of individuals, and customs of working together. It must be remembered, however, that both energy and populations were relatively small. Although military activity was a regular part of early cultures, war was less devastating than it is now.

Whether or not we now regard war and disease as justifiable, they were a regular feature of early preindustrial patterns of human energy support. In this century, we have made major efforts to avoid war, famine, and disease. But regular disease and small-scale, regular warfare may have served the purpose of adjusting populations to resources, thus eliminating the greater ills of large-scale famine and epidemic disease. Primitive cultures were adapted to these ways and accepted them as right; we, of course, have developed different ideas in our recent periods of growth. After we look at our own period in Chapter 10, we can return to the question of regulation of population in the long run.

USE OF STORED ENERGY FOR POPULATION GROWTH

In Chapter 5 (See Figure 5-8) we noted that a large storage of energy can support a temporary burst of growth by a population. After this first growth, the population must decline to a lower level of activity supported by the regular inflows of energy. Whenever human culture changed from one type of energy source to another, it could obtain an extra burst of energy from the leftover storages of the system being displaced. For example, hunting and gathering people of Europe lived off large game animals, to which their populations may have been adjusted for long periods. With the rise of agriculture and the consequent growth of human populations, there were more people to hunt; thus a greater drain was placed on the remaining populations of large animals. Many of the animals eventually became extinct, possibly owing to overhunting. But for a short period the old animal stocks were a special source of energy for transition to a new energy pattern.

Early Agriculture in the United States

The pattern of early American agriculture is an example of growth based on storages. The early pioneers who spread westward over North America in the eighteenth and nineteenth centuries based their growth on the substitution of more intensive agricultural practices for the less intensive practices of the American Indians. The Indians were a mixture of hunters and gatherers and agriculturists. Their populations were relatively low. The pioneers had firearms and other aids that were brought from the Old World—not made from energy produced in America. Using these special products of energy processing, the pioneer displaced the Indians, who depended on their old system.

At first, much of the growth of the new system was taken from the storages of the older ecosystem. Thus, pioneers built cabins and heated them from forest products. Their crops were good at first because of nutrients stored in the soil, where they were being recycled by the forests and prairies. After a few years, these stored nutrients were carried off within harvests, or eroded away because vegetation had been stripped away for much of the year, leaving bare land exposed to rains. The pioneers then moved further west and repeated their pattern. While it lasted, the new system of agriculture had more power than the old hunting-gathering-agricultural system because it was running on storages. Temporarily, more food was produced; but this greater production could not be sustained without additional energy.

The pioneer system started out like shifting agriculture, but instead of going back to used lands in rotation, the pioneers moved on to other sources of energy. They were supported by storages of the forests long enough for them to find additional sources of energy from coal and oil. Through industrial processes, fossil fuels were used to replace the storages disappearing from the environment. People stopped using forests as the source of all energy. Urban factories made fertilizer, and soils were managed with machinery and chemicals. The rapid growth of the United States first developed from energy storages of the wild ecosystems. This pattern was discarded, and agriculture continued on the basis of fossil fuels, yielding more high-quality food and fiber than could have been grown on the basis of the sun alone. The American pattern was quite different from the steady-state agricultural pattern of Asia, where the nutrients were recycled and most of the yields were reinvested in work back on the land. The pattern in Asia was based almost entirely on solar energy.

The situation in colonial America is shown in Figure 9-8, where the storages of a previous system are shown as being used during colonization to develop a new pattern of growth. Without new sources of energy from fossil fuels, growth would have stopped as soon as the soils and forests had been depleted. A new way of life would have arisen; it could have been a lower-energy steady state. For example, after colonization an agricultural steady state developed in Puerto Rico during the nineteenth century. This was more

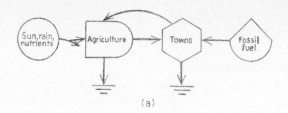

(a)

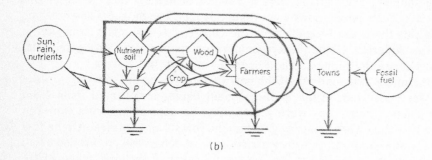

(b)

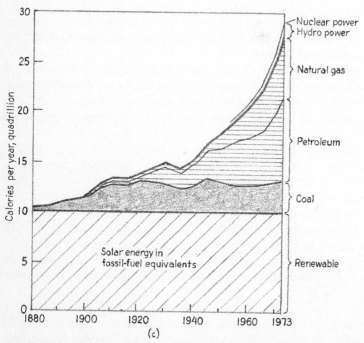

(c)

Figure 9-8 American colonial system first grew on solar energy and initial soil and wood storages, but later shifted to fossil fuels for further growth. (*a*) Summary diagram. (*b*) Energy flows. (*c*) Energy consumption in the United States. *(U.S. Bureau of Mines.)* The contribution of solar energy as a renewable, free, energy inflow is usually taken for granted. Its value in Calories was divided by 2,000 to obtain an approximate value in FFEs.

or less self-sustaining before outside influences began to accelerate population growth. In actuality, however, any steady states developing in the Western world were short-lived. When fossil fuels began to be used in increasing quantities, growth was accelerated, as Figure 9-8c shows. In the harsh conditions of a frontier, a new, growth-oriented way of life developed out of a mixture of people and cultures. In the process six generations of Americans were attuned to a "growth ethic," as we will describe in Chapter 10.

PREINDUSTRIAL TOWNS AS SYMBIOTIC CONSUMERS OF RURAL ENERGY FLOWS

The flows of energy in a town that serves an agricultural area without much fossil fuel are shown in Figure 9-9. Money from the town flows out to the

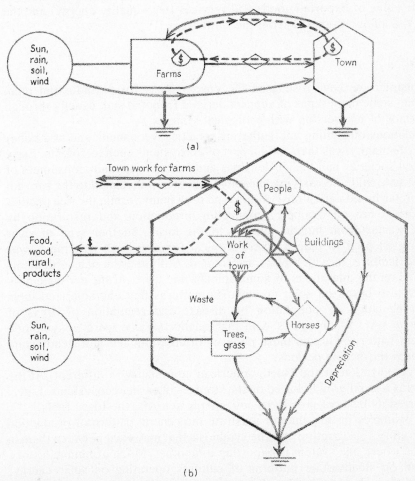

Figure 9-9 Energy basis for a town serving an agricultural economy. (a) Relationship of farms and town: symbiosis. (b) Details on energy flows in the town.

surroundings, where it buys an inflow of food, wood, and other rural products. This flow is transported by horses, processed by people, and made possible by buildings. Hence, the diagram shows all three contributing work to the interaction symbol summarizing the town's overall activity. The work of the town includes maintenance of the people, the buildings, and the horses. Wastes are partly absorbed by the trees and grass, since the concentration of horses and pavements is not large. Horses are shown getting some food from the grass. Products and services over and beyond what is necessary to keep the town running are sold back to the countryside to keep up the money storages. The energy flows of the town as shown in Figure 9-9*b* are very similar to those for any other consumer (see Figure 3-2). There are interactions between the energy storages in the consumer unit that feed back to interact with the inflowing resources from outside. Some feedback of work also goes back outside in exchange for the inflowing energy. The exchange of money helps to match the value of exported goods and services (high-quality energy) and the value of the inflow of energy from the rural areas.[1]

SUMMARY

In this chapter we have discussed the energy basis for preindustrial man, considering some main types of support derived from the sun, usually through various kinds of partnership with ecological systems.

Populations of hunting and gathering societies were small, and as a minor consumer humanity was then truly a part of ecosystems such as forests. Early agricultural societies had higher densities and became the main consumers of the ecosystem. Still, the carrying capacity of an area was related to the sun, not to fuels. Solar-based agriculture of this type took many forms; the rice paddies of the Orient are an example, connecting production and respiration by recycling wastes from human villages to the farm. Another type of solar agriculture, adapted to places where tropical forest growth was rapid, was shifting agriculture, in which plots are rotated. Still other human settlements were based on the migrations of some animals and fishes of the ecosystems— which are also based on solar energy. The patterns that characterized them included regulation of population by disease and regulation of areas of influence by war. When towns with high-quality service were in symbiotic equilibrium with rural areas, these patterns often continued in a steady state without growth for long periods.

The colonization of the North American continent was different. At the outset it was a rapid growth based on the storages of earlier ecosystems. Later, it transferred its basis of growth to new energy sources, the fossil fuels.

The examples in this chapter illustrate the general pattern of production and respiratory consumption and the symbiosis that may exist between them in local areas where one system is exporting to another. Much of human history was based on steady-state patterns of cultures operating on solar energy, sometimes through agriculture and sometimes through other ecosystems from which man obtained inputs and to which he returned services.

FOOTNOTES

1   The following explanation suggests that flows of energy are a measure of value and are ultimately responsible for the values human beings attribute to money. The town isolated in an agrarian region in a steady state (Figure 9-9a) helps us understand how energy flows represent value for survival and why circulating money helps keep track of their values.

In Figure 9-9a money circulates from rural farms to the town and back. The money flowing in one direction is balanced, on the average, by the return flow. The money flows as a countercurrent to energy flow. It starts as low-quality energy in the country and then is transformed and concentrated by production processes and the transport of products to town. In the town, the energy is concentrated further in high-quality goods and services that are returned to the country. For a regional system to compete well and thus survive, it must use its energies in the least wasteful way while generating the most energy flow. To be effective, the energy used in concentrating energy in processes going to the right must be supplied by increased effects on production by the return services back to the country on the left. When such an energy arrangement exists, it is an equal-value loop. The circulation of money around that loop is the way human beings recognize that the flows are of equal value. The fossil-fuel equivalents of the energy flows around the loop are also equal, since both were generated ultimately from the same energy source by successive upgrading processes. The energy diagrams give us a theory of energy value. In steady-state conditions, fossil-fuel equivalents can be used as a measure of the contribution of energy to survival and thus to value. Fossil-fuel equivalents can be used to estimate value in flows that do not involve money for comparison with flows that do involve money.

Chapter 10

Industrial Revolution and Urban Growth

In the nineteenth century throughout the Western world heat engines (see Figure 8-3) were applied to human work—first to ships, trains, and water pumps, and later to other activities—at accelerating rates. Fuels for engines were found and applied to more and more processes. Various mechanical and electrical inventions made it possible to hook complex processing and manufacturing jobs to the rotating wheels of heat engines. The engines became better and more reliable; soon, energy was also being transmitted in electrical lines. Use of energy per person began to rise as the whole system was boosted with fuel engines. The countries in which these rapid developments were taking place were said to be in an *industrial revolution.*

The energy of one person's work was much increased when that person had machines to operate. Even agriculture began to use more and more motorized equipment to manage the land, to harvest, to fertilize, to remove weeds, and to process foods. The energy being added was high-quality energy.

Wherever a person was still doing work by hand alone, in competition with a person using machines, he was at a disadvantage and was soon put out of competition. His service was worth less, and he was paid less. Consequently, his portion of the wealth of his society went down. To earn as much as those

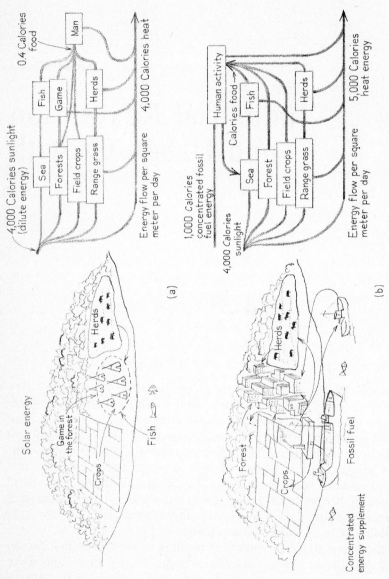

Figure 10-1 Comparison of (*a*) an all-solar economy and (*b*) a fossil-fuel economy. (*Odum, 1971.*)

using machines, people using old ways had to change. Often this meant migrating to the towns and cities where fuel-based work was centered. Work was concentrated where fuel and other energy materials were delivered, often at points of ready transportation. Figure 10-1 compares a solar-based economy with a fossil-fuel economy. Note the many differences.

Energy flowing as coal and oil made the cities places of concentration of many kinds of work which were organized into new industries. Energy, which had formerly come from solar processes through agricultural systems, more and more began to come from fuels through cities. Even agricultural production began to use machinery based on fossil fuels. Some of this machinery worked in cities to produce fertilizer, farm equipment, and chemicals; some of it worked directly on farms. Industrialized farms have now become an extension of the city: many farm workers live in the city and commute to the farm by vehicle. Some main flows of energy into towns and industrialized cities are compared in Figure 10-2.

ENERGY CHARACTERISTICS OF CITIES BASED ON FOSSIL FUELS

More Concentrated Flows of Energy

A new kind of human settlement, the urban development, began to emerge. Figure 10-2b shows the main process in an urban area operating mainly on fossil fuels. Generally, such areas have a high concentration of energy per acre, with more machines, cars, high-rise buildings, computers, complexity, wastes, and flows of money. As long as fossil fuels can be bought cheaply, such cities, with their greater energy flows, win out in competition with simpler towns. People find jobs in the cities, where energy is flowing into industries, government activities, and even welfare. The kinds of storages required in the work

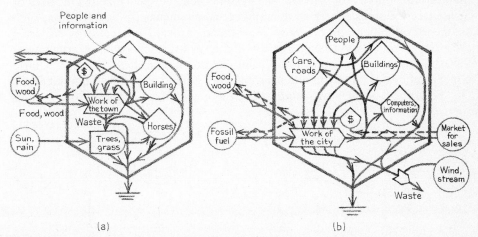

Figure 10-2 Energy-flow model for cities. (a) Town running on surrounding agriculture based on solar energy. (b) Urban area running mainly on fossil fuels.

process of the city are similar to those in the simpler agrarian towns. In the city, however, trees and grass may become much reduced as the density of activity increases. Now wastes are not absorbed and recycled within the city but must be exported and dealt with in waste-disposal plants.

Relationship between City and Farm

Figure 10-3 diagrams the relationship between rural farms and the town (or city). Notice the circular flow of goods and services: money circulates in the opposite direction from goods and services, in exchange for them. The details within the hexagonal units are those shown in Figure 10-2. Figure 10-3a shows dilute solar energy being concentrated on the farm into high-grade food, wood, and fiber that support the town. In the town these products go into making even higher-grade energy flows such as the labor of human beings which produces things to go back to the farm. This diagram is based entirely on solar energy; in Figure 10-3b, on the other hand, there is some fossil-fuel energy. Concentrated energy runs the cities that support the farms. The farms are no longer getting all their energy from the sun, rain, winds, and soils. The second pattern uses both fuels and solar energy: the two interact, with the high-quality energy being amplified by the solar energy.

Mechanisms Stimulating the Economy

In Chapter 5 (Figure 5-6), we described competitive exclusion. When energies are available for growth, alternative systems compete for the inflow of energy, and one system may drive out another. For example, some businesses drive others into bankruptcy, and some cultures replace other cultures. Using coal and oil as resources for rapid urban growth, the countries of the Western world developed characteristics that differentiated them from their competitors. Western cities became centers of growth-promoting activities that accelerated the use of resources for growth. The systems became specialized in the uses of new energy sources. There were many economic means for acceleration, such as the development of business investment, stocks, tax patterns favoring reinvestment, and advertising. The use of energy for transportation, commerce, the stimulation of organizations, and so on, had a high priority. Dense concentrations of people in cities during work hours increased communication.

Figure 5-5 shows two kinds of growth stimulus. B shows the regular

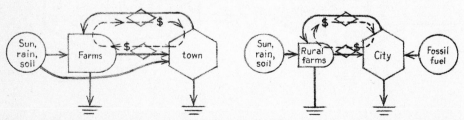

Figure 10-3 Human agriculture coupled to towns and cities: (a) without fossil fuel; (b) based on fossil fuel interacting with solar energy.

feedback from storage to drive inflows that is found in all systems. *A* shows the more accelerated feedback accompanying a more rapid rate of growth attributable to the cooperative interaction of stored assets, including people, buildings, and money. Notice that for this special acceleration a self-interaction symbol is shown (this symbol was explained in Chapter 5). The special acceleration drains more energy and can thus be very negative if efforts to get that energy are unsuccessful.

Belief in Growth

Perhaps even more important than these characteristics was the general belief which developed in our culture, and which has been passed down from parents to children, that growth is good. It was accepted that expansion activities, such as those mentioned above, were good in themselves, even though they had wasteful aspects. Our culture now has many growth-oriented goals, described as *progress* and *expansion,* that are almost taken for granted.

CHARACTERISTICS OF GROWTH ECONOMIES

Profit from Growth

When a farm or business is operated, some goods and services are bought, and some goods and services are sold. In the process work is done, using the combined energies of the fuels that are also bought and the available natural forms of energy—land, wind, water, soil, sun, etc. If the business is successful, sources of energy are tapped to develop enough goods and services of value so that as much money is obtained from sales as is spent in purchases. (Purchases include enough to operate the owner and his family.) If sales are enough to keep the business operating and to pay for repairs, maintenance, replacements, and all hidden costs, the business survives and is in a steady state. If, however, there is enough energy to develop more assets than those required for replacement and to sell more goods and services than are necessary to pay for purchases, then the assets grow and the money on hand can grow. The increase in money that accompanies the growth of assets we call *profit.* Profit occurs only when the energy situation is favorable.

As was shown in Figure 6-9*b*, accumulated energy assets must go into new efforts to pump in more energy or to make the processing of energy more effective if the system is to compete well with others like it. Money profit and the assets it represents, therefore, are used to generate new pathways of work.

Figure 10-4 shows that assets may increase if energies are rich and if production exceeds depreciation and work outflows. On the large scale of a whole nation, such an increase in assets would allow more money to be added without changing the ratio of assets to money supply. The new money is profit.

In a small business, if the new assets are sold, the money accumulated is called *profit.* On a large or a small scale, the new money must be spent and circulated to help keep energy flowing in and to keep the system running.

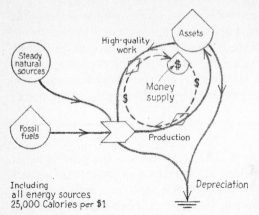

Figure 10-4 Circulation of money in relation to the energy loop in the United States. To keep money equivalent to assets, the supply of money must be adjusted according to the growth or decline of assets.

Net Growth of Assets, Capital, and Investment

Assets, as we have just mentioned, are the results of productive work. Assets include buildings, goods, foods, knowledge, and all other things of value that we save. When the production of assets exceeds losses due to depreciation, the saved assets grow—that is, the storage of assets becomes larger. The extra assets accumulated may be used to obtain new energy; the assets interact with available new energy sources to further accelerate their flow. This pattern, with the feedback loop, has been shown in Figures 3-2, 4-2, and 5-5.

Where money is circulating, it is as a countercurrent to the flow of work, as was described in Chapter 4: money is exchanged for goods and services. The flow of money corresponds to the flow of energy that it releases. For example, in Figure 10-4, 25,000 Calories of production flow through the production process for $1 circulating. Notice the closed circle shown as a countercurrent (dashed line). If inflowing energy is enough to cause production to increase, the amount of stored assets increases, the feedback action of the assets pumps in more production, and the money circulating represents more true work being done. In this situation more money can be put into circulation without changing the amount of work a dollar buys. New money can foster new growth when there is plenty of energy to use.

Figure 10-5 shows assets being transferred in an attempt to tap a new energy source: this is *investment*. Investment can be arranged by borrowing money; money available from loans is called *capital*. What happens to money capital during periods of increasing production, accumulation of assets, and new investment?

The supply of capital money that one can add to stimulate production should depend on the expansion of production and assets with new energy. Extra capital is not available if there has been no previous expansion in production and assets. Investment in new production will not be successful unless new energy is available.

When a loan is made, money is generated. The lender and the borrower

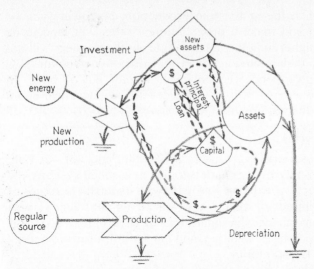

Figure 10-5 Money for investment is borrowed from capital in the main economy to start a new energy inflow by purchasing assets to start a new production cycle. Example: An agricultural economy uses capital and assets to invest in contributions to the economy based on fossil fuel.

both believe they have the value of money in their control. If new energy comes in with the new activity made possible by the loan, there is energy for the increased money.

If, however, there is a steady state and production and the level of assets are constant, then adding money merely reduces the amount of goods that a dollar buys. The true buying power of money is decreased. That is, adding money has caused inflation. The amount of money that can be circulated without a loss in the value of money depends on the amount of productivity and assets being maintained. The buying power of money cannot be expanded unless the energy base is expanded.

Figure 10-5 separates old production assets and circulating money from new production developed by borrowing money to invest in a new source of energy. Transfer of money from an old to a new activity allows the purchase of initial goods and services from old assets to start new assets. If new energy is successfully drawn in, there will be enough new assets to create new money that can be paid back by the borrower, as interest for the use of money. Money will not be loaned unless interest is paid for its use. Interest is really an attempt to add new money.

If the interest paid for new money (a loan) is not accompanied by an equivalent increase in true work (productivity), then the amount of work represented by money will drop, and the true buying power of money will drop by this amount. This is inflation. If the drop in buying power is more than the interest paid, the lender gets back less than he loaned; of course, people will then avoid lending if they can do better by putting their money to other uses. If interest rates are higher than the increase in production, borrowers will be unable to pay interest and still make a profit. Interest rates will tend to be adjusted by lenders to be higher than the inflation rate, so that lenders will not lose the buying power of their money. A loan that does not generate new energy can cause bankruptcy, since the borrower cannot pay it back. In general, if there are no new energy sources to be developed, there will be no new capital and no new industrial growth.

More Complex Economic Cycles

As long as energy reserves were large, easily accessible, and not yet fully tapped, growth was limited only by lags in developing new social systems and money systems. Many aspects of explosive growth, with its unstable surges of money and its uneven distribution of wealth, have stirred our consciences. As a result, there have been many changes in institutions of government, our ways of managing money, our systems of taxation, and our attitudes toward energy and money during the last two centuries.

One of the worst problems during the period of expansion has been the oscillation between times of prosperity, when people spend their money and thus cause new energy flows, and periods of depression, when less money circulates. These are called *business cycles.* To stretch their earnings during bad times people hold on to their money, causing a slowdown in spending which in turn causes a slowdown in the pumping of energy. Economists have learned to stimulate faltering economies by adding money so as to increase spending; and adding money does accelerate spending, even though some inflation is produced.

Consumer Goods and Leisure

Because energy was easily available, and because the industrial economies were ahead of their competitors, energy did not seem important. It was believed that any spending was as good as any other spending, as long as the money flowed. The beliefs developed that the advertising of luxuries was as good for the economy as putting energy into productive processes that could increase net energy. Some consumer products, such as automobiles, did save people's time for more effective work. But the goal of working less and having more leisure time ultimately reduced productivity more than consumer goods increased it.

Control of Disease

When rich new sources of fossil fuels were used, there was growth of assets, production, depreciation, and flow of money. Economies expanded. Because more assets were being added, population could also grow without a loss of assets per person. At first, this expansion of populations caused disease. In the nineteenth century, as people became more concentrated in towns, epidemics developed. Some diseases were spread because of the proximity of people to each other—germs were easily transmitted. Other diseases were set off because sewage got into water supplies. Displaced people who moved to the towns were susceptible to disease because they were coming into contact with new germs. Some epidemic diseases were transmitted by animals—malaria and yellow fever by mosquitoes, bubonic plague by rats and lice.

But since energy and assets were increasing, it was possible to divert energy into understanding and preventing disease and controlling epidemics. Pasteur and others developed vaccines and public health measures to interrupt the old system of controlling disease. Sewage plants were developed to break the cycle by which germs in wastes controlled human populations when they became too dense. Window screens and other measures of mosquito control broke the cycle of insect-borne diseases. By the middle of the twentieth century, many of the epidemic diseases were sufficiently understood that they could be controlled whenever there were enough medical resources and social stability so that prevention could be organized. Epidemics were mainly local, following wars and catastrophes such as earthquakes.

Exploding Population

Not only did the new means of controlling disease stop the epidemics that had been generated by new concentrations of people; length of life was also extended. Expanding populations helped some countries, like the United States, to expand energies further. The new population growth provided more workers for further expansions of energy and more growth. Whatever seemed successful in maintaining an effective economy and making the culture more powerful was regarded as good. Earlier patterns—those of stable societies— had been adapted to low levels of disease, some infant mortality, and the early death of the aged. But an emerging new medical and social ethic emphasized reducing all disease, extending life as much as possible, and expanding the population.

The new ethic; and the accompanying scientific knowledge of preventing and controlling disease, was spread throughout the world with zeal. Missionaries of church groups, government aid programs, and even conquering armies spread the new ethic into countries which were not a part of the industrial revolution. The new ideas reached highly populated countries like India long before the system of using fossil fuels. As a result, population grew much faster than the energy base of the economy; the standards of living went down; and resources and food supplies were not enough.

Over the world as a whole, however, the system seemed to be working. A pattern of life spread which was characterized by a higher consumption of energy per person and a greater role for human beings in the ecology, through machines working on their behalf. Many people in industrialized countries could work fewer hours. They earned enough money for necessities and luxuries, and had choices in their expenditure of time, leisure, and the new wealth.

Fading Consciousness of Energy

Before the industrial revolution people who farmed recognized the value of natural energies. They knew that their productivity depended on the sun, rain, weather, soils, and so on. These people who lived on their farms also realized that human work was necessary to productivity. Because they could see the effects of natural energies and knew the effects of each person's work, they believed that nature and work were valuable. When people with these ideas about energy moved to the towns and cities as the industrial revolution progressed, they carried their commonsense understanding with them. Their work and thriftiness made the early industrial period effective in competition. Cultures which included their values and resources along with the new sources of energy grew much faster than other cultures. They took over much of the resources and power of the world from cultures running on older patterns and still adapted to steady, dilute solar energies.

Soon, the industrialized economic systems were so far ahead of their competition that their decisions about use of excess resources and assets, leisure time, and new energies did not matter. At first, waste did not affect their status, since only they had the machine technology developed to use the new fossil fuels. A whole generation or more did not learn the value of natural energies. They thought more energy was always going to be available. They thought that new technology could do anything without limitations imposed by the availability of energy.

As the cities became larger, the great flows of power in industrial plants, power plants, and industrial farms were far removed from most people, who were at the consumer end of the chains. They had little idea of energy other than that they bought it at a filling station for their cars or plugged in their appliances to electric sockets. The money they paid for gasoline and electricity was small, because it paid only for the work of delivery, not for the true energy of the fuels and power flows delivered. Energy was no longer regarded in a realistic way.

Numerous Sources of Energy

By 1970 the development of human beings and their cities added many dimensions to the production of assets. For example, consider a coastal region of Florida with several sources shown in Figure 10-6. Part of the region's work is done by sun, winds, tides, and waves as in primitive times. Much has been added by economic development that uses income to purchase and bring in fuels and natural gas. Power plants have become important. The modern city uses all available energy. This is another example of the principle that in surviving systems high-quality and low-quality energies interact to do more work.

Power Plants

Associated with the industrial revolution and the growth of cities was the development of technology to generate and use electrical power. Electricity is a

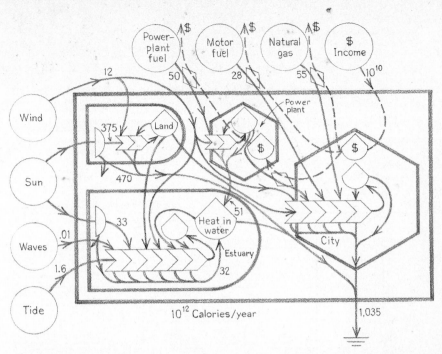

Figure 10-6 Energy flows in northwest Florida in the area served by the Florida Power Corporation. Energy flows of sales and purchases of goods and services are not included in this diagram. *(Odum, 1974.)*

high-quality energy. To generate it, much fossil fuel is required, directly (as fuel) and indirectly (in the construction of complex power installations and transmission lines, which contain much steel and copper). Electrical power is clean and may be used to accelerate a wide variety of activities. Developing electricity and making it generally available made possible more effective coupling of rich energy resources to all the activities of urban life.

The system of developing electricity from fossil fuel is diagramed in Figure 10-7. Notice the feedback of energy required to get and transport oil and then to maintain installations, finally yielding electricity. Almost 4 Calories of oil are required to deliver 1 Calorie of electric power. Oil energy is transformed into electricity by means of a heat engine that turns an electrical generator. (See Figure 8-3 for the energy details of a heat engine.)

ENVIRONMENTAL STRESS

Figure 6-1 showed the high-quality energy flows of man (on the right) depending on the lower-quality energy flow of the sun (on the left). Where energy is supplied from the low-quality end (the sun), the amount of dependent high-quality energy that can be maintained in symbiotic balance is relatively small. The carrying capacity for human beings (a high-quality activity) is small.

With the industrial revolution, a separate source of high-quality energy,

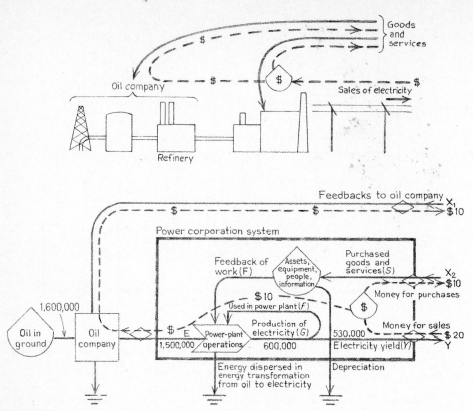

Figure 10-7 Energy flows in an electric power-plant system which obtains needed goods and services by selling electricity. Its operation depends on the ability to purchase oil cheaply. Net production of the system is Y minus X_1 plus X_2. Numbers are Calories of heat equivalents for the flow of one barrel of oil which costs $10.

fossil fuel, was introduced. This made it possible to increase the ratio of high-quality to low-quality activities. The carrying capacity for the complex life of human beings was increased.

Systems using high-quality energy get more work out of their energy when it interacts with, and is amplified by, low-quality energy. This was shown in Figure 6-2. Thus the flows of fossil fuel directly and indirectly diverted energy from the earlier agrarian energy chain to the new system. In centers of urban growth such as the United States, Japan, and Western Europe, natural areas were taken over by interactions between the urban world and solar energy. These interactions included agriculture, industrialized forestry, motorized fishing, use of natural winds and water flows as waste receivers, suburban housing, and tourism. With so much fossil fuel flowing, the amount of low-quality energy available for interaction with high-quality energy became scarce. The uses of the environment to help get more work out of urban investments began to affect essential environmental systems in many places. The symptoms were pollution and other threats to health and increases in the cost of protecting the environment. There were fewer areas where low-quality solar energy could be found to make the use of fossil fuel effective.

A measure of the interaction of urban work based on fossil fuel with solar energy is the ratio of the two expressed in fossil-fuel equivalents. In the United States, where damage to the environment first became a public issue, the ratio in 1973 was 2.5 Calories of fossil fuel to 1.0 Calorie of solar energy (in Calories FFE). The model of energy interaction in the United States shown in Figures 10-4 and 4-2 indicates how high-quality goods and fuel must be matched by low-quality solar energy. In the world at large, the ratio in 1973 was 0.3 Calorie of fossil fuel to 1 Calorie of solar energy.

Those areas that rapidly developed high urban densities have found themselves competing with less developed areas, which have a better match of solar energy and fossil-fuel energy. The areas which make the best match can sell products and services most cheaply. As fuels for growth become less available to those whose initial growth once gave them an edge in competition, energy tends to be redistributed more broadly and evenly. The need to match high-quality energy with low-quality energy creates a limit to urban growth. A region's solar energy ultimately determines its use of fossil fuels.

SUMMARY

In this chapter we examined how systems supporting human beings changed as fossil fuel was applied, first as an auxiliary source and later as a major source of energy. With heat engines allowing almost all activities to be accelerated, the rich energy of coal and oil caused extremely rapid growth. Cultures with traits that accelerated growth—such as profit and capitalism, urban development, and the reduction of disease and prolonging of life by medicine—became dominant. They used the major part of the new energy. Soon their power was being expressed in military and economic activity and in the development of cities. Growth based on fossil fuel was prevailing in the world, spreading from the industrialized countries to the others. The basis for humanity had shifted: human beings were sharing the world with the machines that were their servants and their masters. Specialization developed, and the new culture became used to progress, growth, change and near-miracles of technology, always comsuming more and more fossil fuels. The quality of energy processing was increased by the greater use of electric power, which is very flexible.

In urban areas the increased high-quality human activity based on fossil fuel diverted low-quality energy from the earth's life-support system and produced environmental stresses. In these areas, a shortage of low-quality solar energy to amplify high-quality energy limited growth, with the result that other areas have gained economic advantages.

In this chapter we noted that regional flows of energy support human beings, their power plants, and their cities. We examined how earnings, profits, and loans work in starting new enterprises, and how energy must be available for new enterprises to succeed. In order to understand the changes coming, we looked at the distant and recent past.

Alternative Sources of Energy.

Chapter 11

Processing Energy
Sources for Humanity

Primitive societies, as relatively low energy users, lived within ecological systems supported by the steady energy flows of air, ocean, and earth. Modern cultures have built many kinds of special systems for tapping the energy flows of the earth, diverting them from their former ways of maintaining human life to new ways that go more directly to human beings. Greater use of energy permits more human activity but also makes human beings more responsible for managing the energy of the earth for survival. Human beings have whole systems of industrial activity associated with getting, processing, and using coal, oil, the sun (through agriculture), winds, volcanic heat, nuclear energy, and many other forms of energy. Because all these energy systems involve large amounts of goods, services, and materials and require the use of energy to get energy, we are often confused about which sources of energy are rich and which actually take more energy than they yield. Many uses of energy pollute the environment. In terms of energy, pollution interferes with the energy flows of the environmental life-support system. When the effect on nature's support is taken into account, the overall effect of an energy user's system can be negative.

In Chapter 6 we used the energy diagrams to show hidden relationships and introduced the idea of net energy. In this chapter we will examine the systems for gathering the main kinds of energy that support modern societies, and some systems that have been proposed as potentially useful. We must be aware of the net energy, if any, of each type of energy flow into our culture. We want to know which sources will be of use in the future, considering the relative amounts of net energy they may yield. Many issues of public policy, and many alternatives for putting our tax money into searches for energy, depend on understanding the net energy of various existing and proposed developments. An understanding of our sources of energy is also necessary to the consideration of the energy crisis and inflation; these we will take up in Part Three. Figure 11-1 shows the main flows of energy into the economic and

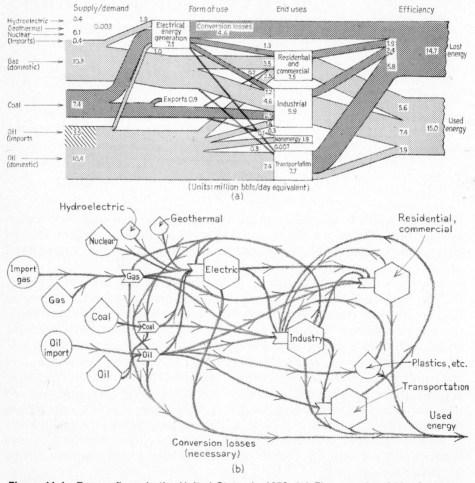

Figure 11-1 Energy flows in the United States in 1970. (a) Flow graph; width of pathway indicates size of flow. The term *lost energy* is misleading: energy is not lost but goes into heat as a necessary cost of upgrading the quality of energy. *(Joint Committee on Atomic Energy, 1973.)* (b) Energy diagram.

industrial activity of the United States in 1970. Most of the main energy flow was fossil fuel (coal and oil); this went into transportation, industry, commerce, and housing, and into power plants to make electricity.

Let us now consider these various sources separately, with the hidden flows by which one source aids another and by which environmental energy flows often exert unrecognized support.

COAL

Figure 11-2 shows coal mining. Coal is mined either by using machines to strip off overlying rock or by tunneling when the coal is deeper. Mining requires energy to operate machines and indirectly to operate that sector of the economy which provides goods, services, and labor. When land is disturbed by mining, the production processes of natural vegetation or agriculture are interrupted until the land has been revegetated. If this is left to nature, many years are required. But energy may be used to regrade and reseed the land with productive vegetation so that production is restored sooner, thus increasing the overall energy contribution of the vegetation. Energy diagrams help us learn which procedure gives more energy.

Coal is shipped by barge or rail to be used in power plants, industry, or domestic heating. Pipelines that carry water slurries of coal are now being used as an alternative to railroads. Transportation is expensive in terms of energy—not only directly, in the motors involved, but indirectly, in the development of steel and other products required for rails and pipelines.

Consideration of coal from the point of view of energy requires that we include the use of the coal. This creates some by-products, such as ash from burning, which must be dispersed. Some coals have a high sulfur content (2 to 3 percent); this comes out of the smokestacks of power plants as acid vapor (acid volatiles are shown in Figure 8-9). Acid vapors wash some fertilizer elements out of soils, strain human lungs, and dissolve some architectural materials (such as marble). These effects are part of the "energy impact" of coal.

Whereas coal is very concentrated energy, the net effect of using it may be negative if the deposits of coal are deep in rock or dilute. The feedback energy costs of mining are large, and the negative environmental effects are large. There may be so much energy used or diverted in mining and processing coal far below the surface that these activities produce a net loss. Very deep coal is best left in place; as part of the normal earth cycle, it will contribute to long-range geological work or come to the surface for easier use later. There are still large reserves with net energy, but less than the quantities quoted to the public by those who do not make net-energy calculations for the deep beds.

Figure 11-2b is a simplified summary of the energy yields and costs for mining 1 million tons of coal from the Great Plains region of the western United States. If only the energy costs of mines are considered, the yield at the mine is 425 FFEs, compared with feedback costs of 10 FFEs and environmental costs of 0.7 FFEs. The yield ratio is very high—40 to 1. If, however, we consider the energy costs of transporting the coal to a city in another state and distributing

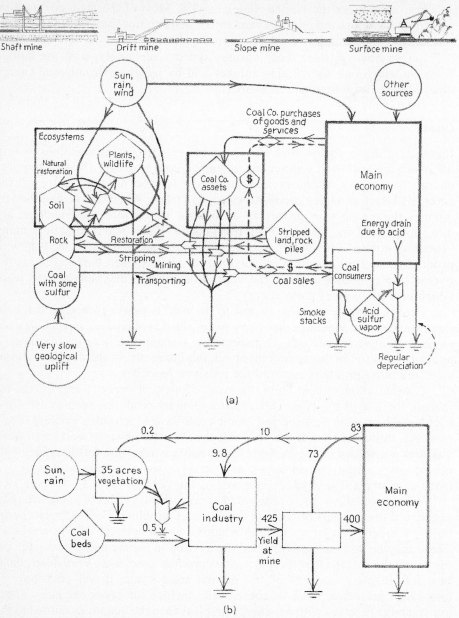

Figure 11-2 Energy flows in processing coal by strip mining; environmental effects are included. (*a*) Detailed energy diagram. (*b*) Summary with energy-flow estimates given for 1 million tons of coal. *(Ballentine, 1975.)* Energy costs include transportation to a Midwestern city and distribution in a power plant. Numbers are 10^{10} Calories of fossil-fuel equivalents.

power through an electric power plant, the additional energy cost is 83 FFEs and the yield ratio becomes less—4.8 to 1.

Figure 11-2*b* illustrates an important principle in calculating net energy, that the amount depends on where you are in the energy chain. The net energy supplied to the city through the power plants has a lower ratio than purchasing Arab oil and transporting it to a coastal city, and thus at present prices the coal may not be as competitive for these cities. However, if people move to the place where the coal is being mined and set up cities there, the net energy will be much greater. Will this be the trend? Which kind of use should be encouraged, that which uses existing cities, that which moves people to new sources of energy, or a more conservative use by people who have moved from the cities to the farms? Which makes the economy more competitive?

Figure 11-2*b* also shows that the environmental damage is substantial but much smaller than the energy yielded. The energy cost of restoring the land is very small compared with the energy yield. Requiring this restoration is not a large energy cost compared with the yield. The environmental stress per year is large enough that it pays to restore the land: the energy saved by restoring plant production sooner is larger than the energy cost of restoration.

GAS

Organic matter buried in sediments undergoes changes that convert some of it into organic gases; we call these, collectively, *natural gas.* Natural gas contains the fuel gases hydrogen, methane, butane, and propane, and other gases in various percentages. Since the sedimentary deposits are under pressure from the weight of the land that covered them over, the gas normally flows out on its own, so that it is very cheap to process. Of all the fossil fuels, natural gas is easiest to process and easiest to transport in pipelines, and it burns hottest. Because it is easily used, and because its price has been held down by government regulation, it is being used up first in the United States. Natural gas, the most valuable energy, may be the first of the fossil fuels to become unavailable at any price. See Figure 11-3.

It has been suggested that we buy some components of natural gas from foreign countries, cool them until they liquefy, and import them in refrigerated ships. This involves some danger: a failure of refrigeration, or an accident, could release clouds of flammable gas which might drift over towns, causing explosions or causing suffocation by displacing oxygen. Railroad transport of propane is an example of this kind of shipping; there have been some propane explosions, when railroad cars were involved in accidents.

Some American towns have made their own gas from other fuels, for the convenience of having gas to operate stoves and refrigerators. Gas, such as methane, can be made in small quantities from sewage and from some kinds of garbage, or as a by-product of making coke and charcoal. Making coal into gas is a process being considered; this would transform coal into a fuel that could be sent by pipe more easily.

Networks of pipelines connect many parts of the country with Texas,

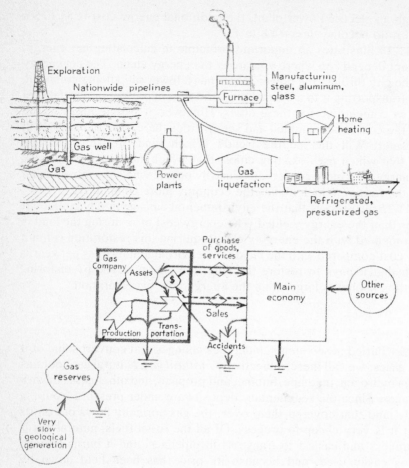

Figure 11-3   Energy flows in processing natural gas.

Oklahoma, and other areas where natural gas was abundant. These pipes might
eventually be used to transport other kinds of fuel, such as artificially made gas,
slurries of coal, or the heavier fossil fuels (for example, certain grades of oil).
Human ingenuity in putting old things to new kinds of uses has no limit, so long
as there is some energy source with which to do this. Artificial gas, it should be
noted, is likely to yield less net energy than the rich natural gas that is now
mainly gone in the United States.

OIL

Figure 11-4 shows oil-processing systems. Oil, being lighter than water, drifts
upward from its original sedimentary deposits. It collects under any kind of
geological rock layers which are pocket-shaped and not porous (Figure 11-4*a*).
In general, oil is found most abundantly in pockets that occur above deep
layers of oil-forming sediments. These are deposits where the burying process

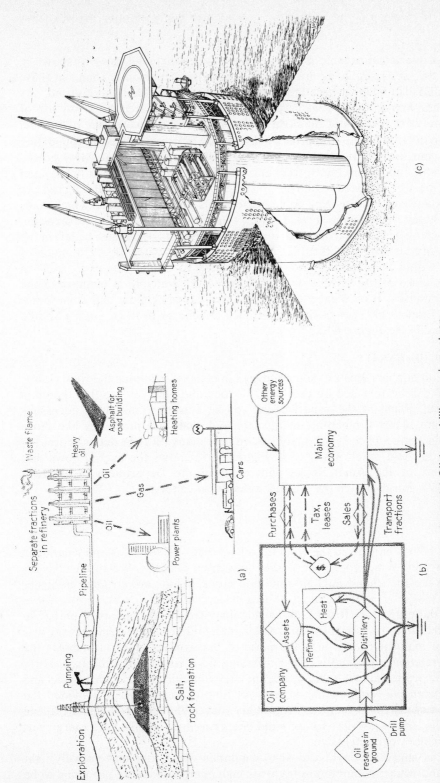

Figure 11-4 Oil. (a) Oil processing. (b) Energy flows. (c) Offshore drilling equipment, an example of the enormous storage and cost of high-quality energy needed to gain more inaccessible energy. (*Tetra Tech, Inc., 1973, "The Effect of Natural Phenomena on OCS Gas and Oil Development," prepared for the Council on Environmental Quality under contract no. EQ4AC010.*)

covered over algal organic matter before it was consumed. As was shown in Figure 8-7, sediment beds are deep at the shore. More oil is found in these areas, and also inland where the sediments are being folded and uplifted.

When human economies first shifted from other sources to heavy use of oil, there were pockets of oil close to the surface; these were easy to find and cheap to drill and pump. But after a half-century of use, oil wells are found only deeper and deeper. In 1973, the average well was 1 mile deep. "Dry holes" (failures to find oil) are more frequent, and very large amounts of money and energy must be invested to find new oil (Figure 11-4b). Offshore drilling rigs are very expensive because they contain enormous amounts of steel, which takes much energy to make and form (Figure 11-4c). The saltwater causes corrosion, storms cause a great deal of damage, and large amounts of oil must be left in the ground because of the difficulties of operating underwater. By 1970, much of the oil within the United States that was easy to obtain was used up. Rich sources close to the surface were, however, still available in quantity in the Arab countries, Iran, Venezuela, Nigeria, the Soviet Union, and some other places. Even at 1973 prices there was much net energy to those buying Middle Eastern oil (see Figure 6-8).

Oil Rock (Oil Shale)

Not all oil bubbles upward; some remains in sediments among the rock grains. This oil can be released by crushing the rock and heating it so that the oil is vaporized. Some of the vaporized oil can be burned to make the necessary heat. Pilot plants for getting energy from oil shale use processes like those shown in Figure 11-5. It is not yet known how many oil-rock deposits are rich enough to yield much net energy. Figure 11-5c shows the energy flows that were part of the Anvil Rocks pilot plant of the Department of Interior, 1945–1975. Far more energy was used than was yielded.

WIND

Human beings have used the wind for a long time—in sailboats and windmills, for example. Wind cools our bodies and dries laundry. We often take for granted the fact that winds air-condition cities. Winds are stronger in higher latitudes.

In the era before fossil fuels, windmills were major sources of energy in the Netherlands, in North Dakota, and in many other places. Wind energy was used for many purposes in many parts of the world. Figure 11-6 shows the energy relationships involved in harnessing wind for mechanical uses. Early windmills were made of bulky wood, but later windmills were made of steel and other expensive, high-quality inputs. As long as solar energy provided the wood, or cheap fossil fuel was the primary source of energy for steel in electric windmills, it was possible to use wind energy to do work. The net energy was often small or even negative.

As is suggested in Figure 11-6, windmills which run on relatively light winds and whose manufacture involves high energy costs are not going to be

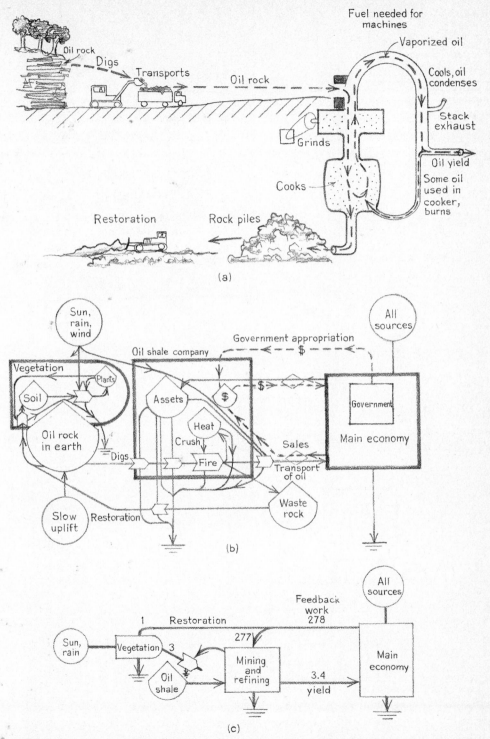

Figure 11-5 Energy flows in processing oil shale. (*a*) Simplified sketch of the process. (*b*) Energy flows. (*c*) Summary of energy flows in Anvil Rocks pilot plant, 1947–1975. (*Gardner, 1975.*) Numbers are 10^{10} Calories of fossil-fuel equivalents.

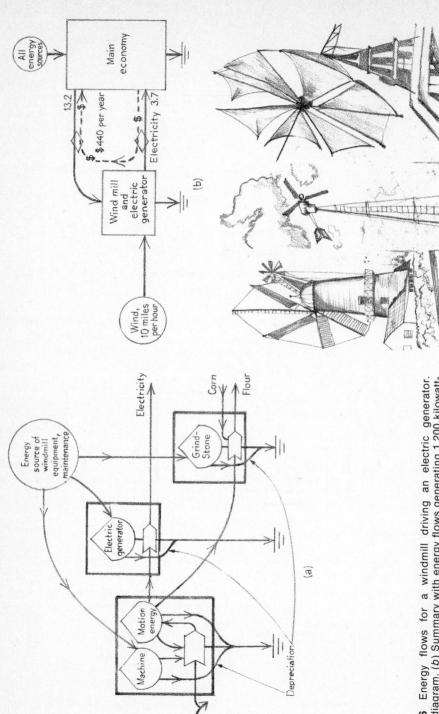

Figure 11-6 Energy flows for a windmill driving an electric generator. (a) Energy diagram. (b) Summary with energy flows generating 1,200 kilowatt-hours of electricity. Numbers are millions of Calories of fossil-fuel equivalents. (c) Windmills.

much help in the future. As costs of other energies that were used to subsidize windmills increase, the deficit in net energy will increase further, making the technologically complicated windmills less and less practical.

TIDES AND WAVES

The pulls of the moon and sun generate tides; that is, they cause the surface of the oceans to rise and fall once or twice a day. These changes cause the ocean to surge in and out of bays and estuaries, doing work on the ecosystems and sea bottoms in those areas. We are already using the energy of the tides when we eat seafood from estuaries and enjoy the beauty of our coastlines. We also time the sailing of large vessels to coincide with tides.

Harnessing tidal energy for electrical power has been attempted in some places where tides are large. Tides are largest in the mid-latitudes; in some bays where tidal actions converge, the water may rise and fall 20 feet or more. For example, at Rance in France the estuary is dammed and water is allowed in and out only through tubes with turbines that generate electricity. The yield ratio is 12, a high figure. Whether the removal of this energy from the Rance estuary has decreased estuarine productivity is not known.

Winds keep the sea agitated by waves that develop continuously and run across the sea to break on beaches and other shores. The action of waves at the shore does useful work in maintaining beaches and keeping them clean. Waves also operate a circulation of water through the sand that helps keep seawater clean. It is estimated that the waves of the sea filter the oceans' waters through the beaches of the world in 1,000 years. Thus we are already using wave energy. Harnessing it for other uses has not been done very much. Enough energy might be harnessed to be important in some localities, such as islands, but not enough energy could be tapped to operate cities.

HYDROELECTRIC POWER

Elevated water falling downward from mountains does work as part of ecosystems and the land cycle. Energy of elevated water usually yields net energy. Many of the world's rivers have now been dammed to divert the energy of elevated water into hydroelectric plants to generate electricity (see Figure 11-7). Once, water was used to turn primitive wooden water wheels; now it turns steel turbines. In both cases, second energy sources are involved. Wood used in water wheels comes from solar energy through forests; steel is a product of fossil-fuel economies that make it cheaply. The net energy of water power will be less when the energy from fossil fuel that supports the general supply of goods, services, and materials is less. Energy of water is of high quality and generates most work when it can amplify other energy flows. People who try to use the last water power of river basins for electricity often forget that the water is already doing important work, supporting agriculture, forests, soils, flood plains, recreation, coastal fisheries, etc. Developing hydro-electric power may involve losing energy that is as valuable as the electricity

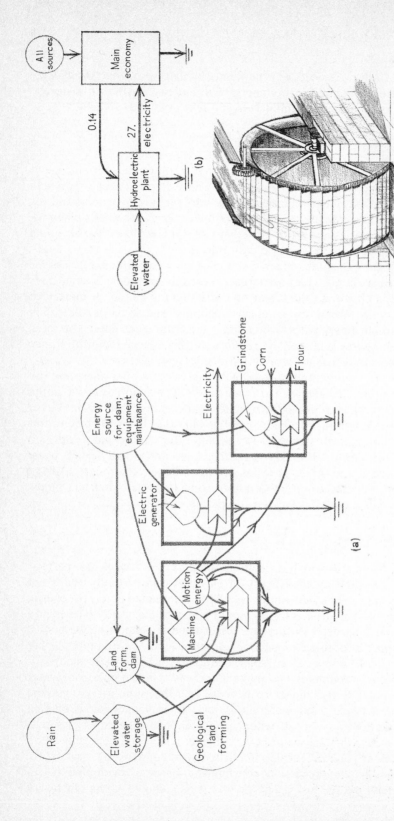

Figure 11-7 Energy flows in processing hydroelectric power, including a simple water wheel and a hydroelectric plant. (a) Energy diagram. (b) Summary of energy flows in hydroelectric plants drawing water from 100-foot dam. (Young and Odum, 1974.) Numbers are millions of Calories of fossil-fuel equivalents per kilowatt of electricity. (c) Water wheel.

produced. The Aswan Dam on the Nile River in Egypt is an example of the controversy over which pattern is more energy-effective. In the old pattern, the Nile annually fertilized and cleansed by flooding, and it supported fisheries. In the new pattern, it supports canals for irrigation and generates electricity for industry—but there are problems, such as the filling of the dam with sediment, and a blood disease transmitted by snails in the canals.

SOLAR ENERGY

Human existence is already based on the solar energy that keeps the atmosphere and seas circulating, makes food, and preserves a range of temperature in which human beings can survive. The initial absorption of sunlight converts light into heat of land or water which then heats the atmosphere. Figure 11-8 shows a "low-energy" use of solar energy, drying clothes.

Many people have been misled about the potentialities of solar energy because the Calories reaching the earth are so large. It must be understood that the *concentration* of energy is small, so that by the time sunlight is concentrated, it does relatively little work when net energy is considered. We have already discussed the very important direct and indirect use of solar energy by human beings on their farms and by forests, seas, and lakes to produce many inputs supporting human life. Agriculture is now mostly based on fossil fuel; it uses relatively little land and thus relatively little of the solar energy available for growing food. Using more land in agriculture is one way of using more solar energy.

Many propositions have been made for converting solar energy more directly without separate steps into higher-grade energy, as in generating electricity or manufacturing chemicals. Such propositions are often impractical. As we discussed in Chapters 6 and 7, systems of vegetation have already developed ways of maximizing the concentration of solar energy from its dilute source to organic matter and ultimately to coal and oil. For many millions of years there has been natural selection for the best system of converting solar energy to chemical energy. The chain of biochemical machinery in plant photosynthesis has already been selected for maximum power.

Solar Technology

Solar water heaters have been used for a long time, and of course glass windows are used to let sunlight into houses. These uses, however, may entail considerable energy costs for glass, plastic, pipes, and insulation. As energy for the required materials becomes more scarce, the technology for solar heating may become less practical. Other devices, such as mirror collectors, photoelectric cells, and glass collectors, do not yield net energy. It is very controversial now whether solar technology saves energy or wastes it. Unlike agriculture, solar technology uses sunlight directly instead of through the food chain of plants.[1]

In Figure 6-7 energy flows for photoelectric cells are given. A tiny area of surface receiving a tiny bit of sunlight requires a very large energy expenditure

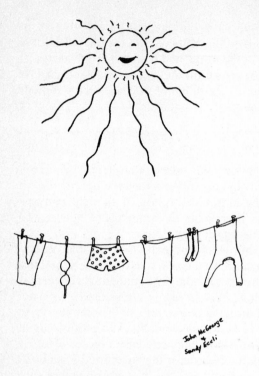

THE ORIGINAL SOLAR DRYER!

Alternative Sources of Energy.

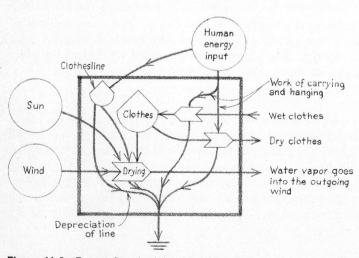

Figure 11-8   Energy flow in doing laundry by hand.

in fossil fuel for manufacture and maintenance. The yield ratio of sunlight harnessed to fossil fuel fed back into the equipment is very small. There is no net energy, and compared with most uses of fossil fuel the energy return is small even for an auxiliary process. Fossil fuel can be converted into electricity more efficiently by being used for power plants than by being used for making photocells. Photocells may be justified for special control purposes, but not as energy sources.

Solar heating of houses and water uses much fossil fuel indirectly in the installations. Solar heaters have been used for a long time in sunny climates. Figure 11-9a and b compares a solar water heater and a fossil-fuel water heater.

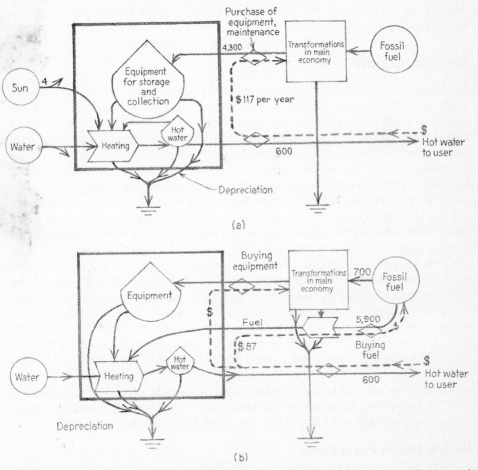

Figure 11-9 Energy flows in solar technology; comparison of (a) solar water heater and (b) fossil-fuel (gas) water heater. Numbers are thousands of Calories per year (fossil-fuel equivalents). (Zuchetto and Brown, 1975.) Both systems use fossil fuel indirectly to supply and maintain equipment. The solar heater takes more equipment but uses less fuel directly. The fossil-fuel heater involves less equipment and storage and less depreciation but requires the purchase of fuel.

In this case solar energy is being used to make low-quality heat energy. In sunny climates, using some sunlight along with fossil fuels, energy is saved by solar heaters as compared with electric and gas water heaters. However, solar heaters use much more fossil fuels than solar energy when they are compared on an equal-quality basis (fossil-fuel equivalents). No net energy is yielded. Capital costs at the start are high in terms of money and energy; savings are obtained later, after the heater has run for several years. One unanswered question is this: Would more energy have been saved if fossil fuels had been put into something else rather than into solar water heaters? As the energy for required materials becomes more costly, technology for solar heating will also become more costly. One concludes that there are ways to conserve energy using solar technology, but solar technology is not a source of energy for running our economy generally.

NUCLEAR ENERGY

Nuclear energy is obtained by operating heat engines using the great concentrations of heat that are generated by reactions in nuclear materials. Nuclear reactions release some of the potential energy of the interior of the atoms of radioactive substances. Of all our sources of energy, nuclear reactions may require the most kinds of special feedbacks of energy. It is very expensive, for example, to make the process safe in the event of accidents, earthquakes, etc. For thirty years, nuclear energy was directly subsidized by fossil fuels. Costly as these installations and processes were, they would have been much more expensive if all the goods and services had not been made with cheap fossil fuel. During this period, much nuclear fuel was mined and stored, research was done, many kinds of reactors were built, and the government incurred huge costs to support the Atomic Energy Commission. Figure 11-10 is a simplified view of the various energy flows involved in atomic energy and some of the ways by which energy subsidy was drawn from the main fossil-fuel economy. Some net energy will eventually result, but the energy cost of getting nuclear energy into usable form is very high. How high is not yet clear. The return for energy invested may not be as large as with other sources; estimates suggest that 2.7 Calories of energy are yielded back for 1 Calorie invested; fossil fuels yield more.

There are several kinds of nuclear processes to be considered as possibilities for the future. Some are in operation now; some may yield enough energy to be used. We will not know until they have been developed further.

Nuclear Fission Power Plants

In 1973, there were fifty nuclear power plants operating in the United States and another fifty under construction (Figure 11-11a). These were fission plants; their operation is diagramed in Figure 11-11b. The nuclear reaction, called *fission,* takes place when rods of uranium fuel excite each other in a reactor

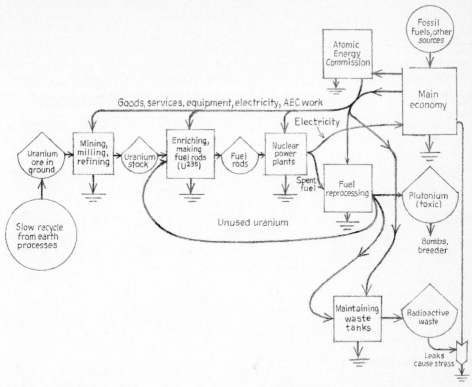

Figure 11-10 Energy flows in the atomic-energy system; note the heavy feedback from the main economy.

tank. It releases huge amounts of very intense heat. This is what takes place in an atom bomb: uranium comes apart in a chain reaction. In the fission plants, the reaction takes place more slowly, however. Fluid circulates through the reactor tank, carrying the heat to a second tank where heat is exchanged across the walls of pipes to create steam. The steam drives the power plant like any other heat engine (see Figure 8-3). Outside water is brought in to carry off the heat that is left over. When this water is discharged, it is a thermal waste; its recooling sometimes adds cost, as for cooling towers. Radioactivity occurs only in the first pipe system, although sometimes it leaks into the steam. Ensuring safety when dealing with radioactivity is very expensive, and this expense reduces the net energy. If the 100 plants produce well for their expected lifetime of thirty-five years to forty years, there will be a net yield of about 2.7 to 1, not counting costs of waste disposal or accidents. A major accident could absorb the net energy in additional costs. Nuclear fission is not expected to be a permanent energy source, because the supply of uranium is limited.

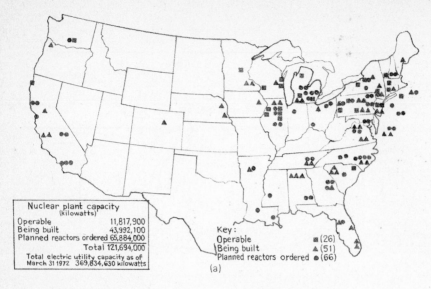

Nuclear plant capacity
(kilowatts)

Operable   11,817,900
Being built   43,992,100
Planned reactors ordered 65,884,000
                  Total 121,694,000
Total electric utility capacity as of
March 31 1972 369,834,630 kilowatts

Key:
Operable   ▣ (26)
Being built   ▲ (51)
Planned reactors ordered ⊛ (66)

(a)

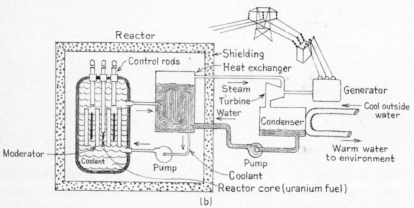

Reactor

Control rods
Shielding
Heat exchanger
Generator
Steam
Turbine
Water
Cool outside
water
Condenser
Moderator
Coolant
Pump
Pump
Warm water
to environment
Coolant
Reactor core (uranium fuel)

(b)

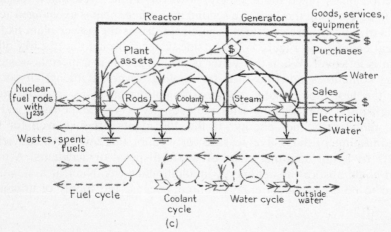

Reactor   Generator   Goods, services,
                                    equipment
Plant
assets $ $
                                    Purchases
Nuclear
fuel rods   Water
with
U²³⁵   Rods Coolant   Steam   Sales
                                    $
                                    Electricity
Wastes, spent   Water
fuels

Fuel cycle   Coolant   Water cycle Outside
                 cycle   water

(c)

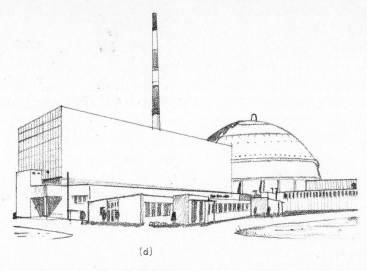

(d)

Figure 11-11   Power plants that use nuclear fission of uranium. *Opposite page*: (*a*) Nuclear power reactors in the United States. (*U.S. Atomic Energy Commission, June 30, 1972.*) (*b*) Regular nuclear-fission reactor. (*U.S. Atomic Energy Commission.*) (*c*) Energy flow in the fission reactor. *Above*: (*d*) Nuclear power plant.

Breeder Reactors

Breeder reactors, such as the one diagramed in Figure 11-12, are arranged so that some of the initial nuclear reaction produces secondary nuclear reactions that create additional radioactive fuels. The radioactive reaction of one kind of fuel causes a second type to become radioactive and capable of reacting to make more heat. Because the reaction generates some new kind of fuel while using the first fuel, it is called a *breeder* reaction. There are some breeder-reactor pilot (test) plants in operation, but how much net energy they will generate is not yet clear. One uncertainty is the costs (high in terms of both money and energy) that will be necessary for long-term storage of radioactive wastes. The breeder processes involve a very poisonous element, plutonium. Because plutonium is so toxic, special care must be exercised in the operation itself, in disposing of wastes, and in preparing for possible accidents. Should there be an occasional accident, the interruption of use of a surrounding area and replacement of facilities would represent a major subtraction from the net energy that the proponents of breeder reactors hope will be generated. The reprocessing step in the cycle has turned out to be very expensive, and no recycling plant is operating regularly. At present, large amounts of money are going into the testing of breeder reactors both in the United States and abroad.

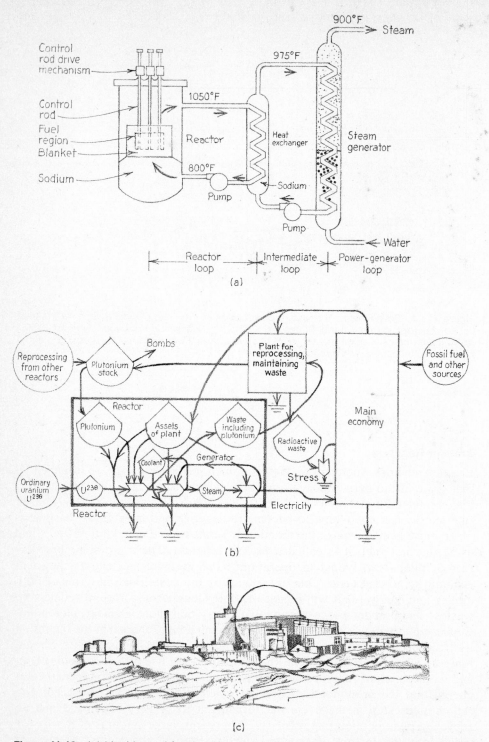

Figure 11-12 (*a*) Liquid-metal fast breeder reactor and fuel reprocessing plant required to complete the breeder cycle. *(Environmental Education Group.)* (*b*) Energy flows in the breeder reactor and proposed recycling process. Plutonium radiation makes more plutonium from ordinary uranium (U^{238}). (*c*) Breeder reactor.

Fusion

The hydrogen bomb involves a different nuclear reaction, called *fusion,* using the element hydrogen. The great goal of many proponents of nuclear energy is to make usable energy from the fusion reaction of the hydrogen bomb. This reaction takes place in the sun, whose mass is so great that gravity holds the very hot gases close enough together for the hydrogen reactions to take place (Figure 11-13*a*), releasing the nuclear energy of the atoms. On earth, however, gravity alone is not enough to hold the reaction together; it blows itself apart, as we observed in tests of hydrogen bombs. Research is going into various ways of holding the hot gases together. Much energy feedback is required for this (Figure 11-13*c*). The temperature is so high that any kind of container would melt. Efforts are being made to contain the reaction with laser beams (Figure 11-13*b*), giant magnets (Figure 11-13*d*), and other means. So far, all the methods

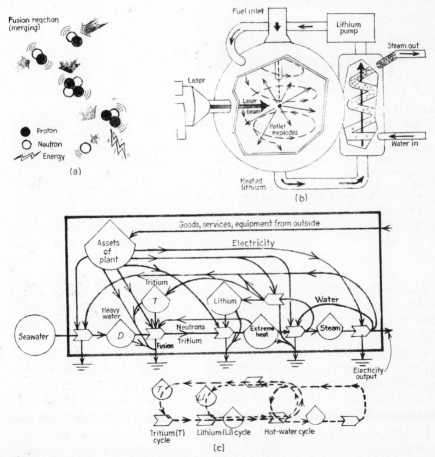

Figure 11-13 Energy flows will have to be evaluated to determine whether fusion yields net energy. (*a*) Fusion reaction. *(U.S. Atomic Energy Commission.)* (*b*) Fusion using lasers. (*c*) Energy diagram of fusion. *(Continued.)*

(d)

Figure 11-13 (Continued) (*d*) Fusion apparatus using magnets. This large and expensive Tokamak-type laboratory apparatus, called an *Ormak-a Diffuse*, is used to experiment with the fusion reaction using a ring of giant magnets to hold the process together; it is about 18 feet tall. If the experimental equipment is so large and expensive, will the apparatus needed for an actual power plant yield net energy? *(Oak Ridge National Laboratory, Oak Ridge, Tennessee, operated by Union Carbide Corporation for the U.S. Energy Research and Development Administration; and* Science, *vol. 172, May 21, 1971, cover, copyright 1971 by the American Association for the Advancement of Science.)*

tried have been fairly complicated and expensive. We suspect that the energy cost of containing the reaction may be higher than the energy released by it.

If fusion should yield net energy, there would be no shortage of fuel, since common materials—such as water—are used as fuel. Some scarce materials are also needed in the reactions, however; for example, lithium is required in one process. It is difficult to draw an energy diagram for a process that is not yet operating on earth; but Figure 11-13*c* will help to show the questions we have about energy costs and subsidies to fusion.

If a large amount of energy is to be obtained from this process or some other process not yet in sight, an even more serious question will arise: whether human beings can sufficiently regulate the biosphere, in the presence of these rich energies, so that our life-support system is not displaced. Such large-scale new energy would operate huge activities that could disrupt lands and seas and displace humanity from the planet. There would be no stopping growth if energy were not limited. See Figure 5-4*a*. The maximum-power principle suggests that no system can stop growing and survive if its competitors have access to untapped energies, for they would overgrow it in competition. Fusion could be disastrous to humanity either if it were so rich that it gave too much energy, or if it took all our capital and gave us no net energy. (See Chapter 16.)

GEOTHERMAL ENERGY

As was discussed in Chapter 8, the processes of the earth's atmosphere, ocean, and crust operate on heat energy where there are differences in temperature. Most of the differences in temperature in natural processes are small. In a few places—for example, where mountains are being built—large differences in temperature develop. Human beings divert energy from the temperature gradients of the earth to run heat engines, but these engines require more temperature difference to operate than large natural processes. Our engines cannot obtain net energy from the small differences in temperature that exist in ordinary earth, air, and water.[2] Tapping the heat of the earth has been economically successful only in the vicinity of volcanoes: steam from volcanic regions is used in California, New Zealand, and Iceland. Figure 11-14 shows the

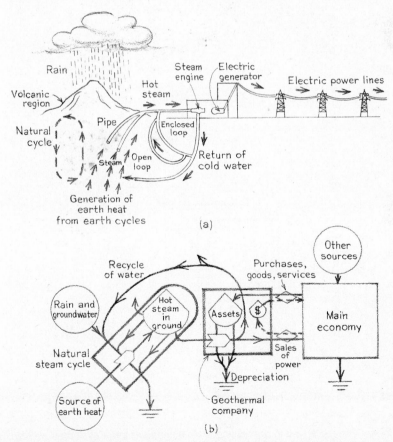

Figure 11-14 (*a*) Geothermal energy process. Open and closed water-circulation loops are shown. (*b*) Energy flows.

use of circulating steam to operate heat engines on geothermal heat sources. However, people could save heat by building their dwellings underground even where earth heat is ordinary.

CRITICAL MATERIALS

Since all materials that have use are themselves sources of energy, it is appropriate to include them in this chapter. We explained in Chapter 2 that all materials carry energy as they interact in chemical reactions and other processes. Scarce, required materials are high-quality energy because they have amplifier effects, but they may not provide much energy by themselves.

Inflows of necessary and critical materials, such as copper and iron, require other energy for processing. The more dilute the critical materials, the

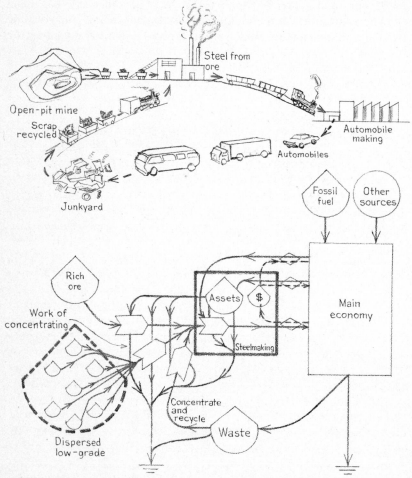

Figure 11-15 Energy flows involved in processing critical materials such as copper, or iron for steel.

more energy is required to process them. As is shown in Figure 11-15, several feedbacks of energy are involved in mining, processing, and recycling critical materials. Like fossil fuels, iron and copper cost more and more energy as the rich deposits are depleted. Since these materials alone do not usually yield net energy, we do not usually think of them as energy sources. The increasing energy cost of getting these materials is another factor in the decrease in total net energy that is causing inflation. When iron and copper deposits were rich and near the surface, we used them cheaply for railways, transportation by automobile, and electric transmission. Now we will have to make do with less of these services, since more energy is required for concentrating and recycling these metals.

ENERGY OF PURE WATER

In addition to the potential energy in elevated water which we use in hydroelectric plants (discussed on page 175), there is potential energy in pure freshwater as compared with seawater (3.5 percent salt) or water which contains dissolved substances. When clean water is used to wash dirt from clothes, wastes from industrial processes, or salts from desert soils, the pure water is an energy source.[3] After the water has dissolved the salts and other substances, it is no longer as pure: it has lost some of the energy that goes with purity. When freshwater reaches the sea, the energy difference between it and the saltwater generates currents in estuaries, does geological work on the sediments, and controls the ecological processes in the mixing zone. The energy of pure water is large, and in states with little mountainous elevation, such as Florida, the energy in the purity of rainwater is higher than the energy in the water's elevation. We are already using the energy of pure water in many ways that we take for granted. It has been suggested that we might harness this energy for power plants, but this has not been attempted on any large scale. Since we are already using this source of energy in our economy (although we do not often call it energy), any such development would divert energy from one need to another.

The energy of water purity is high-quality energy. Freshwater is required for society, so that where it is scarce we use other energy sources to generate it from saltwater. The sun generates freshwater by evaporating seawater, operating weather systems and causing rain over the land. This was diagramed in Chapter 8. Special industrial plants (desalination plants) make freshwater from saltwater using other energy sources to do the same thing as the sun. See Figure 11-16, which shows the fossil-fuel equivalents required to convert seawater to freshwater. Some plants evaporate and condense water the way sun does in nature; others use pressure to force water through membranes that hold the salt back. Whatever the process, there is a large energy requirement for developing pure water from saltwater. The resulting product is a storage of high-quality energy. Because much energy is required to purify water, there is a tendency for human cultures to develop where good sources of freshwater are available.

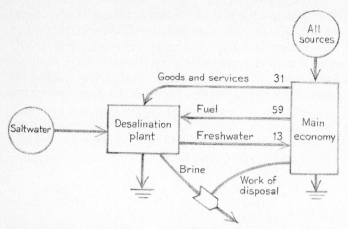

Figure 11-16 Energy flows in converting 1,000 gallons of saltwater to freshwater. Numbers are thousands of Calories of fossil-fuel equivalents.

USE OF HYDROGEN TO TRANSMIT ENERGY

If nuclear energy should become a major yielder of net energy, the problem would arise of sending the energy out from complex fusion plants to users. To do this by means of electricity would worsen the shortage of copper. Also, conversion to high-quality electricity would not be necessary for all uses; indeed, some processes are not readily operated with electricity (imagine farm tractors with long extension cords). One possibility is to make hydrogen gas from water, using the heat of the nuclear plant, and to send it out through gas pipes. Hydrogen gas burns with a very hot flame and is a major part of natural gas. Trucks and automobiles might run on tanks of compressed hydrogen gas. But hydrogen gas is a high-quality energy; and it might not pay to make this gas for all uses. This is not an idea for a source of energy but rather an alternative means of distributing energy if in the future there are rich energy sources that require centralized giant facilities.

SUMMARY

In this chapter we have examined sources of energy that are presently available or proposed. We considered the main systems that yield net energy—such as gas, oil, coal, and nuclear power—and showed their hidden energy subsidies and the ways in which they are taking more and more feedback energy. Some sources, like solar technology, wind, tides, and geothermal energy, were shown to yield relatively small amounts of net energy except in some local areas.

Our aim in this chapter was to simplify each energy inflow system in order to clarify the main factors that make it a source of *net* energy or not. The diagrams presented for the breeder and fusion nuclear processes, when finally evaluated, will indicate whether or not energy from these sources can be large.

Data on these sources, however, are not available, since they are not yet operating on a scale that would provide a true economic test. If we examine various propositions by means of energy diagrams, the faith that rich energy sources will always be found to replace those we have exhausted may be shaken. Certainly, a leveling or decline in energy flows is more likely than further growth, if our discussion in this chapter is correct and no other rich sources emerge.

FOOTNOTES

1 In Chapter 6 we described the theory that energy of one quality is most effective when it interacts with energy of another quality. Another aspect of this theory is that energy of one quality interacts readily with energies that are neither much more concentrated nor much less concentrated. There is, for example, the possibility that the interaction of fossil fuel, a high-quality energy, with solar energy directly without intermediate steps may not maximize energy flow. Too much energy may be lost in such interactions. The system for harnessing solar energy developed in nature involves intermediate steps through plants. Our present use of solar energy in agriculture is successful and also uses the plants.

2 Heat engines operate on the temperature difference between their source and their environment. In power plants the source is the heat of the furnace; in nuclear power plants the source is the heat of the nuclear process. The operating temperature difference within the plant must be maintained so that there will be steam on one side of the turbines and condensed water on the other side to return to form steam again. To condense the steam, it must be passed through cooling coils so that the heat can exchange with the environment by pumping river water, estuary water, or streams of air. In the process the environment is heated. This heat may be harmful to ecosystems not adapted to that temperature. When available, using water for cooling is less costly in terms of goods and services from the main economy than using air-draft cooling towers by which air is blown through the coils.

    Questions often asked are: Why must some heat be released to the environment if heat is energy? Is this release of heat to the environment a waste due to poor design? The answer is that there is no net energy in small temperature differences. When the temperature difference between circulating water and the environment gets down to 14 degrees Fahrenheit or less, the energy costs involved in equipment, goods, and services in getting the additional energy into work is greater than the energy yielded. In other words there is no net energy in the last few degrees of a temperature difference if one uses a power plant. The environmental systems with large sluggish aquatic and atmospheric circulations do convert these smaller temperature differences into water currents and wind which we use.

3 The potential energy in a difference in salt concentration in Calories of heat equivalents is given by the chemical formula $\Delta F = RT \ln (C_2/C_1)$ where ΔF is the energy per mole, R is the gas constant, T is the temperature (in kelvins), and C_2 and C_1 are the salt concentrations before and after energy is used. The energy quality is about three times that of fossil fuel.

Chapter 12

International Flows
of Energy

No part of our planet is entirely isolated, for the biosphere operates as a single energy system. As was discussed in Chapter 8, human life on earth depends on large, worldwide energy systems: the oceans, the atmosphere, and the cycles of the earth. In the last century, worldwide human systems have also developed, including the international economic system, multinational corporations, the balance of military power, global information networks (radio, television, and newspapers, for example), and conscious efforts at international organization (such as the United Nations and international relief agencies). All these systems have emerged in response to the energy sources of the world and have facilitated these flows of energy. Because flows of energy are international, individual countries, states, and cities are dependent on the fortunes of the world as a whole. Much that affects individual lives has to do with worldwide trends in energy: energy systems can, for example, cause economic changes and wars. Consideration of the international flow helps us to see how the existence of our own country is related to that of others. Considerations of energy suggest ways to predict the economic future, minimize war, and eliminate waste in defense spending. Perhaps we can best visualize the way energy controls world affairs if we start with the worldwide distribution of energy sources.

DISTRIBUTION OF ENERGY

As we saw in Chapter 11, human beings in the modern world draw part of their energy from fossil fuels and part from solar energy and its secondary flows—wind, weather, currents, etc. (see Figure 11-1). The main distribution of sunlight is shown in Figure 12-1. In the course of a year, tropical and subtropical regions receive twice the energy of the polar and temperate latitudes, mainly because they have bright sunlight in all seasons, whereas areas nearer the poles are in shadow during the winter.

The winds generated by the difference in temperature between the equator and the poles tend to be greatest where the contrasting temperatures interact—at mid-latitudes. The sun-driven hydrological cycle drops the most rain in the mid-latitude belt of frontal storms (see Figure 8-4) and along the equator, where tropical winds from the two hemispheres come together and rise. The conversion of energy in plant production is greatest in those areas that have a combination of much sunlight and adequate rainfall and nutrients. See Figure 12-2.

The distribution of fossil-fuel reserves by country is shown in Figure 12-3. The Soviet Union and the countries of the Middle East have oil reserves which will allow them to dominate during the next twenty years or so, before these reserves are exhausted. The reserves of uranium for nuclear energy are largest in the western United States, Canada, Sweden, Spain, and South Africa, in that order. Out of an estimated world total of 2 million tons, the United States has about 67 percent.[1]

REGIONAL SPECIALIZATION

Some parts of the world receive large quantities of solar energy. Others have large stores of fossil-fuel energy. Others are in a geographical position that allows them to use secondary, derived energy flows such as strong winds, rivers, and migrating populations of fish. Human beings in different areas specialize, developing patterns of culture and economic activity that generate the most productivity. By emphasizing its best source of energy, an area can maximize its power and its ability to trade for additional energy and products.

In earlier chapters it was suggested that more work is obtained by using high-quality energy flows such as scarce materials, water, and fossil fuels to interact with and help process the lower-quality energy of sunlight than by using each type of energy separately. See Figure 6-2. Different areas of the world receive different combinations of energy flows in different quantities. Therefore, these areas develop different systems for interacting the various energy flows so as to make urban and agricultural productivity most effective. Areas with similar combinations of energy flows may be expected to develop similar systems of energy use. We may expect similarities in culture, occupations, and trade among such areas. For example, different Arctic peoples have similar ways of living, adapted to sharply distinct seasons. Cultures that were formerly different are becoming more alike as their energy basis shifts from

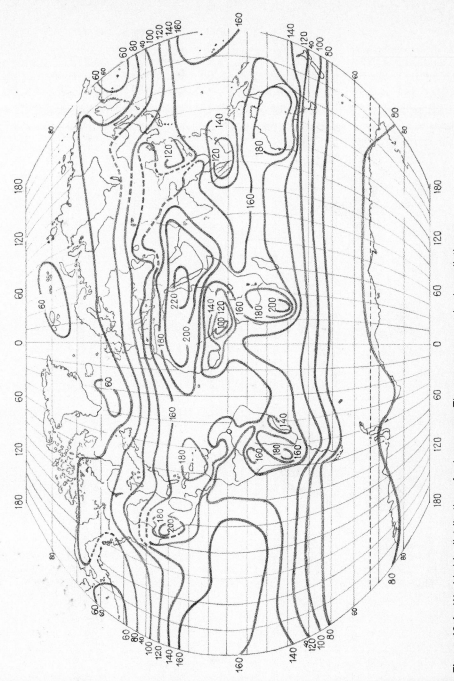

Figure 12-1 Worldwide distribution of solar energy. The average annual solar radiation on a horizontal surface at the ground. The units are Calories per square centimeter per year. (*W. D. Sellers, Physical Climatology, University of Chicago, Chicago, Ill., 1965. After Budyko.*)

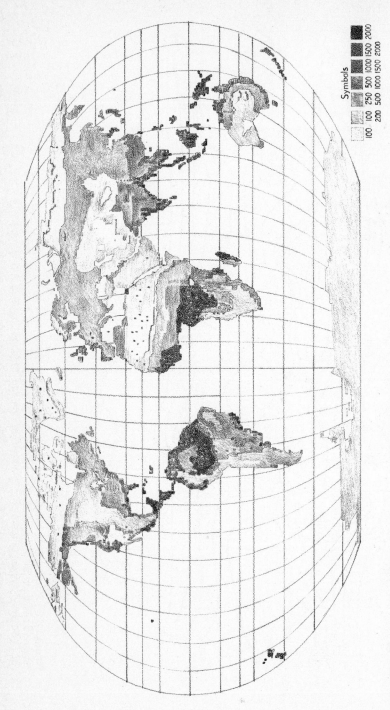

Figure 12-2 World distribution of plant production on land. Annual production of organic matter in grams per square meter per year. *(H. Lieth, "Primary Production: Terrestrial Ecosystems," Human Ecology, vol. 1, no. 4, pp. 303–332, 1973.)*

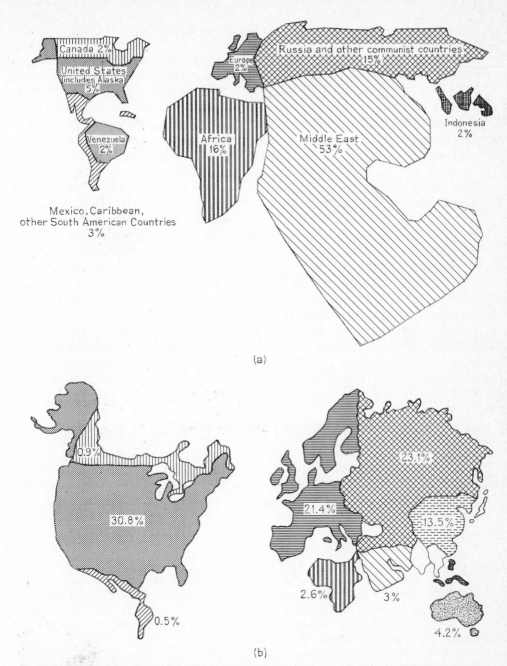

(a)

(b)

Figure 12-3 World distribution of (a) oil reserves and (b) coal reserves. Percentages are shares of the world's known, extractable reserves in each area. The maps show what the world would look like if the size of each country were in proportion to its reserves. *(Exxon Corporation.)*

characteristic patterns of regional support to the general basis of fossil fuel that is similar the world over.

Sometimes experts from one country try to recommend procedures for another country, considering only one process at a time, rather than the whole system of energy interactions. But it may be inadvisable to use methods that work in one energy situation for another, different energy situation. For example, crops that are successful in a country with a developed fossil-fuel agriculture may fail in a country which cannot supply supporting chemicals. Housing in desert regions must obviously differ from that in rainy regions.

Cultures of the world have sometimes been grouped according to their food habits. This gives three main groups: (1) Grain eaters. These are the people who live where the seasons vary sharply—the temperate and monsoon regions—and thus favor the growing of wheat, corn, or rice. (2) Root-crop eaters (root crops include sweet potatoes, casava, manioc, and taro). These people live in climates where the seasons are more uniform, favoring the domestication of plants which store and protect food in roots. For both grain and root-crop cultures, animal food is scarce; only the minimum amount required for balanced nutrition is eaten. (3) Meat eaters. This group includes people dependent on concentrations of migrating animals on land or in the sea, such as the Eskimos and the cattle-herding tribes of Africa. It also includes people with fully developed economies which are based on fossil fuels and use them to enrich agriculture. Meat is higher-quality energy (making it costs a good deal of energy), and these cultures use more than the minimum needed for bodily functioning.

CLASSIFICATION OF NATIONS BY ENERGY ROLE

For some discussions of energy, the nations of the world can be divided into four categories according to storages of energy as fuel and to developed assets. (These are shown, with their energy diagrams, in Figure 12-4). The amount of sunlight received is also important in determining a country's ability to produce, as Figure 12-4 shows. Most countries have substantial solar energy for use in interactions with their high-quality energies (fuels and developed assets). The four categories are as follows:

1   Nations that have developed assets and enough fuels and critical raw materials for their own needs, or more. Examples: the Soviet Union and the United States (until recently).

2   Nations that are industrialized but import fuels and raw materials. Examples: Japan, Germany, the Netherlands.

3   Unindustrialized nations that supply more fuels and critical raw materials than they use. Examples: Saudi Arabia, Iran, Zaire, Nigeria.

4   Nations that are not industrialized and must import fuels and critical materials. Examples: India and Egypt.

Some countries fall between categories, of course. The nations in category 1 are strongest, since they supply all their needs and consequently produce

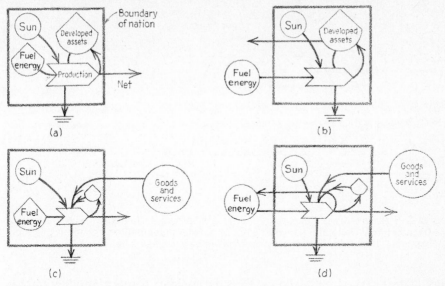

Figure 12-4 Classification of nations by sources of energy (internal or external) and developed assets for production. (*a*) Nations with assets and fuel within their borders. (*b*) Nations with developed assets but which import main fuels. (*c*) Nations with their own energy sources but with undeveloped assets. (*d*) Nations with few assets or energy sources other than the sun.

most. However, nations in category 1 soon move to category 2 and ultimately move to category 4 because of their rapid processing of resources. "Supply nations" (category 3) seem to be moving into category 1.

TRADE

Energy Basis for Trade

A country's ability to trade depends on the energy resources within its borders that can be used to produce an energy flow of special value to another country. Figure 12-5 shows the interaction of fuel and critical raw material reserves, sunlight, and developed assets within a country producing goods and services for export that are sold for money. This money is sent back out to buy needed fuels or goods and services. The more energy sources the country has within its borders, the more cheaply it can sell its goods and services. The more cheaply it sells its products, the more of the market it captures and the larger is its volume of trade. The more it sells, the more additional energy flows it can buy. Having its own energy sources, it can easily purchase additional flows. A country is helped if the energy stimulus of what it receives is greater than the energy cost of what it trades.

Trade and World Productivity

In discussing the productivity of a farm, we saw that the flow of one type of energy can interact with a second type to release higher-quality energy such as is required for the support of human beings. Food, clothing, and housing

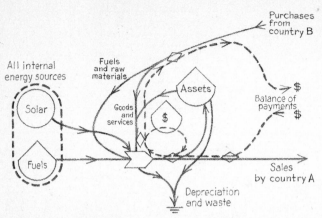

Figure 12-5 Relationship of sources of energy and pathways of energy loss to balance of payments in external trade and hence to a nation's general competitive economic situation. Balance is more favorable when energy sources are better and waste is less.

require many kinds of energy flows. Thus the complex forms of production that support human beings are maximized by the bringing together of the various outputs of energy-specialized regions. Diversity of energy flows can generate additional energy. Exchange and world trade can increase flows of energy by eliminating special shortages in some areas. It often takes less energy to exchange for some necessity than to use one's own energy to supply it by some process at home. If, on the other hand, the energy cost of swapping and transportation is higher than the energy gain to be made by the exchange, then supplying the necessity locally is better. When fossil fuel was harnessed to heat engines, transportation was one of the first activities to be accelerated. World trade was facilitated, and fossil fuel was used to eliminate shortages that were restricting productive energy flows. Rich fossil fuel favored organizing activities on a large scale.

Balance of Payments, Energy, and Money

One of the most talked-about principles in international economics is the need for a "balance of payments" in trade. This can be considered either in terms of energy or in terms of money. If country A buys $1 million worth of goods and services from country B, then country A must get $1 million worth back by selling, directly or indirectly, enough goods and services to country B. The balance of payments for one country is shown in Figure 12-5. The balance of trade for two countries is shown in Figure 12-6. Figure 12-7 shows more detail, including the circulation of money, which should balance if trade is to continue. In Figure 12-6 and 12-7 country A is shown as having more sunlight and less fuel resources than country B, so that their trade can be complementary.

A balance of money payments is required because there must be a balance in the exchange of energy. To understand exchange, consider Figures 12-6 and 12-7. In Figure 12-6, two countries are trading their special products. Suppose

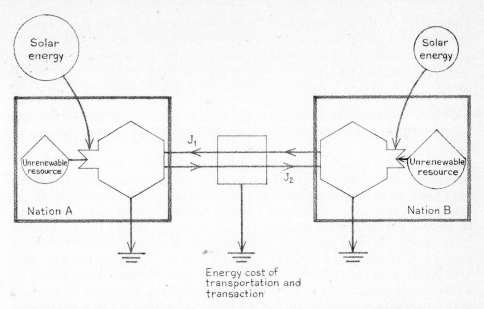

Figure 12-6 Energy flows between two nations with international exchange. The surviving pattern develops a balance of payments of energy contribution to work. An energy balance of payments exists when the energy effect of J_1 equals J_2.

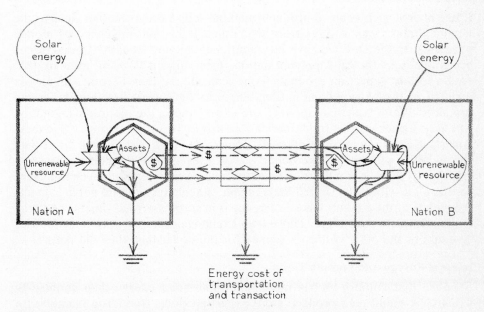

Figure 12-7 Money flows between two nations with international exchange.

that country A is exchanging food for machinery from country B. Country A must get for its exchange as much of a contribution to its own productivity as it exports, or more. If it gets more, it can store the difference and use it for growth, diversification, or new exchange, as was discussed in Chapter 6. If it gets less in trade than it sends out, it will start to deplete its own storages: that is, it will lose energy by this kind of trade. In this case, using production internally is preferable to using it externally. Economies that prevail and survive will be those that maximize power by maintaining their balance of payments.

To balance payments of energy, there must be a balance of money payments. Most people try to think in terms of money because their individual lives seem to be dominated by it; therefore, let us examine Figure 12-7 to see how the energy balance generates a money balance. Notice that the exchange of the two energy flows is made by means of money. Money from country A buys machinery from country B, and money from country B buys food from country A. If country A buys $1 million worth of machinery but gets only $500,000 from the sale of food, it will not have enough money to buy any more of the machinery it needs to make the food. Country A must raise the prices it charges for goods so that it will get back its money. Country B would have to spend all the money it received to get its necessary food at the higher prices. If there were a third country, selling food more cheaply, country A would have to drop out of this particular international activity and do something else to get foreign exchange.

Devaluation of Money

If one nation, such as an oil-producing nation, sells more to another, such as the United States, than it buys from that other, it has an imbalance of money payments. In the case we have just mentioned, money from the United States accumulates in the oil-supplying nation faster than it is being used by that nation to buy American products. American money then becomes an excess quantity in the oil-supplying nation, where its only value is to buy American products. While American money is in excess there, the oil-producing nation will pay higher prices for American products in order to turn its excess American currency into useful goods. Money will tend to flow back faster than the energy it purchases. The dollar is worth less. Since the dollar represents less energy flow, it is inflated. The oil-producing nation will then charge more for the oil it sells to the United States again. When payments are not in balance, the money of the country whose currency is in surplus abroad—in this case, the United States—loses value. Those who exchange currency between countries give less of the other country's money for the dollar than they did before.

Internal Energy and Foreign Trade

Suppose that country A has more renewable solar energy than country B. Country A could then produce more export products (food, for example) for foreign exchange than country B. Country A could lower its prices, and more money would flow in. The accumulation of foreign money would make it easy to buy more products from country B and thus get more energy from abroad. The country with the best internal energy sources, therefore, tends to have a favorable balance of payments of energy and money.

The country with less internal energy must export more of what it produces to get what it needs, so that it develops less assets. It becomes economically disadvantaged, sending the yield of its work to further increase the assets of the country with the larger internal energy base. The United States used to have more internal energy than other nations and thus had a favorable balance of payments, was able to grow, accumulated assets, and developed concentrated high-quality activities at home. As American sources of energy began to be used up and other nations processed more energy and could charge high prices for it, our balance of payments began to be unfavorable. Growth and the development of assets and high-quality activity shifted to the energy-supplying nations.

Gold

Because it does not corrode and is scarce, gold has always been highly valued. It has been used as money in many countries at many times in history. Even when one country was conquered by another and its money system was replaced, inflated, or declared worthless, its gold coinage was regarded as valuable in itself. In modern times, the quantity of money flowing has become so large, and the institutions for handling paper money sufficiently trustworthy, that gold has become less important than money. However, when money is losing value because of inflation, having gold is sometimes regarded as a way to keep from losing one's savings. Since less gold is available than is needed for this purpose, gold is open to speculation, and its market price has gone up and up, possibly far above its inherent value in the energy system. As true energy becomes scarce, and more people try to convert gold to meet basic needs, its price may fall back to reflect its ordinary value as a product useful in jewelry, goldenware, and industry.

CONCENTRATIONS OF POPULATION AND ENERGY RESOURCES

When public-health measures and modern medicine developed, these practices were exported to all countries of the world to help people. Modern medical treatments were spread throughout the world by missionaries, private foundations, and governmental programs. These had the effect of eliminating the previous systems of control of population by disease. This, along with expectations about the standard of living and the possibility of growth, caused an acceleration of the rate of population growth in many countries (Figure 12-8).

Some of these countries, like France and the United States, started with relatively low concentrations of population and were able to increase education and fossil-fuel technology quickly. For the individual, conditions were im-

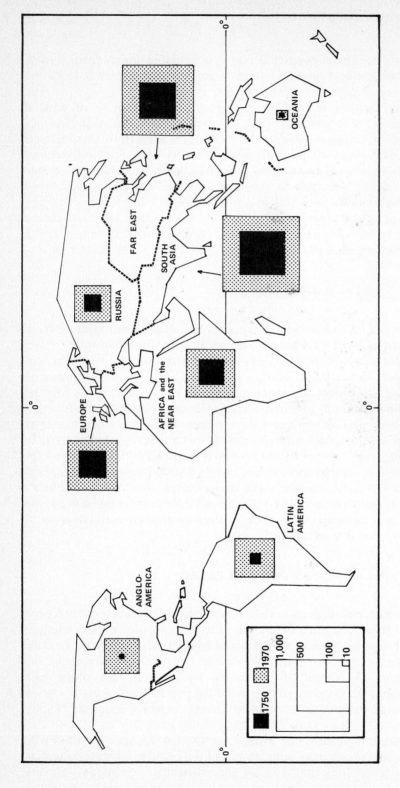

WORLD POPULATION
1750 - 1970

Figure 12-8 Population distribution before and after the industrial revolution. Numbers are millions of people. *(S. McCune, "The Population Explosion and Its Effect on the Environment," Proceedings of 1971 DuPont Environmental Engineering Seminar, bull. ser. 137, Engineering and Industrial Experiment Station, University of Florida, 1971, pp. 1–16.)*

proved and the average energy expended on behalf of each person increased. In industrialized countries machines replaced many people's jobs. Methods of birth control were developed, and attitudes towards large families changed. Population began to level off before the energy leveled off. Energy per person could remain more or less steady.

However, in countries like India and Egypt that started with large populations and had not developed much fossil-fuel technology, even larger populations developed and the amount of energy expended in support of each person was very low. These countries have tried to industrialize and to increase their energy support, but it is doubtful if this is possible without first decreasing population. When shortages of fossil fuel and consequent rises in prices developed, it became difficult for these countries to import much energy. Energy per person became even less.

Distributing energy equally among people has been suggested as an ideal solution; but this would sometimes be contrary to economic vitality and the maximization of power. Energy must first be used to get more energy and to make the use of energy efficient. Distributing energy equally among people will not help if the work of the people is unequal or if there are more people than are needed to make a system compete successfully. Considering the world as a whole, energy and people are in different places. World population growth is given in Figures 12-8 and 12-9b.

OCEANIC FOOD RESOURCES

The oceans of the world were once regarded as outside national boundaries. They were free, since they seemed to be both inexhaustible and uncontrollable. But as human control of energy increased, it became possible for more and more of the sea to have its fish harvested, its bottoms mined, and its coastal areas subjected to control by governments. As Figure 12-9a shows, much of the most productive marine ecosystems are along coasts. This is because concentrations of nutrients and favorable water motions develop there.

Fisheries are rich in zones of "upwelling water." Where water circulates up to the surface from below, it brings mineral nutrients which accelerate productivity (see Chapter 8). Production is stimulated by the interaction of land and water, runoff and tide, wind and continent. Most of the fisheries are located where the productivity of the sea is concentrated, in the relatively thin layer of water above coastal shallows. The shelf of land that surrounds most continents extends out gradually from a depth of 50 to 500 feet before the bottom drops off to the ocean floor; this is called the *continental shelf*. This shelf of land is geologically a part of the continent and part of the zone of sedimentary depositing shown in Figure 8-7. Most of the best fisheries for shrimp, lobster, king crab, sardines, and bottom fishes are found over the continental shelf near the shore. These fish are high-quality food, rich in protein.

Figure 12-9a shows zones of fish yield in the sea. They include shallow shore zones and zones of upwelling. Geologically (as has been noted) and

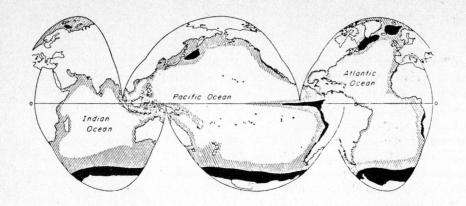

(a)

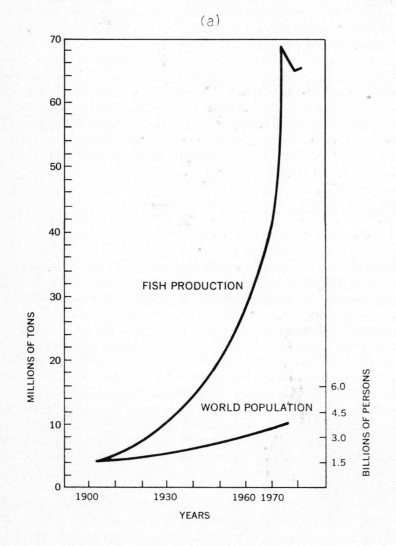

(b)

biologically, these shelves have some relation to the countries they border, and in recent years efforts have been made to include them within national boundaries. Instead of patrolling 3 miles out, as was traditional, many nations now go out 300 miles, even in areas where there is no shelf. Fossil fuels have made possible large factory ships with subsidiary small fishing vessels that can go to the shores of other countries for fish.

Much of the productivity of the sea that reaches the edible food chain is now being caught, and many species are being overfished. *Overfishing* means catching so many members of a species that not enough are left to breed and grow for the next season's catch. Some fisheries are so overfished that their existence is threatened: whales, for example, are in danger of extinction. Many specialists believe that the steeply rising curve of world fish production (Figure 12-9*b*) has reached its peak, because there are few stocks not already being fished to the maximum. World fish yields have not increased in the last few years.

Very high energy from one system can threaten the existence of other sources of energy if one is used to exhaust the others without a feedback of energy (see Chapter 3). Figure 4-5 showed harvesting of fisheries without feedback to maintain them. What feedback should human beings be making to keep the fisheries that they prefer competing with other organisms in the sea? An open question is: Should the fisheries be managed by an international organization, or should the sea be divided and the fisheries placed under the jurisdiction of nearby countries? If rich energy should become less available, then no one may have the energy either to use the sea fully or to dominate its management. Overfishing may then decrease.

WORLD FOOD POOLS

During periods of expanding energy, technological agriculture, and modern medicine, the world as a whole produced enough food and at the same time expanded populations. By the 1970s there was doubt that food production could be expanded much more. Agriculture is now based mainly on fossil fuels, and fossil-fuel energy is being rapidly depleted and becoming harder to get. Also, many of the methods of stimulating agriculture have been used, and now yields increase less for each effort to stimulate them. When countries with too many people for their own food supplies have experienced famines, the rest of the world has sent its surpluses, preventing the terrible episodes of starvation

Figure 12-9 World fish production. (a) Distribution of zones of fertile mineral nutrients over the world's oceans. The black areas are the richest, the white the poorest. *(Paul M. Fye, Arthur E. Maxwell, Kenneth O. Emery, and Bostwick H. Ketchum, "Ocean Science and Marine Resource," in Edmund A. Gullion, ed.,* Uses of the Seas, © *1968 The American Assembly. Reprinted by permission of Prentice-Hall, Inc., Englewood Cliffs, N. J.)* (b) Growth in fish production and world population compared. *(Updated from* Fisheries of North America, *issued by the North American Fisheries Conference, April 30–May 5, 1965, sponsored by National Fisheries Institute, Inc., Washington, D.C.)*

*Reproduced by permission of Gib Crockett and
the Washington Star-News.*

and disease that once accompanied famine. No really strong force has
compelled any country to limit population or to change its culture to avoid
famine. By the 1970s the world food pools existed only in the industrialized
agricultural nations, and these pools have begun to disappear as costs of fossil
fuels rise. Any further growth of population is sure to bring worse food
shortages and more famines.

These changes in quantity and distribution of food are likely to shift the
responsibility for food supplies back to regions. Nations will find their own best
ways for producing food and will depend less on food which must be brought
from outside and is produced by international food technology based on fossil
fuel. Land devoted to agriculture will be increased; fewer machines and
fertilizers will be available.

Exchange of Wheat for Oil

Countries with sufficient sunlight and rainfall can exchange wheat or other food
for oil from countries like the Soviet Union and the Mideast nations whose
agricultural production is less. If the United States exchanges 50 million tons of
wheat for 620 million barrels of oil, the yield to the United States, expressed in
fossil-fuel equivalents, is 5 Calories for 1 Calorie sent abroad and 1 Calorie
used in the agricultural work. Industrialized agriculture in the United States can
continue temporarily as long as energy inputs can be sustained.

TERRITORY

In primitive tribes and among many species of animals, the group or individual defines a territory and defends it against groups or individuals in neighboring territories. There is some fighting at boundaries, which establishes the distance at which those in the center can be effective. Where energy resources for a unit are high, its territory can be larger.

Something like this may be involved in power politics between nations. Energy-rich nations can dominate larger areas than energy-poor nations, since they have more energy for growth, military defense, and long-distance transportation, communication, and organization. When the energy available to a nation changes, its ability to maintain its influence—economically, militarily, and culturally—weakens, and it must contract.

This may have been a factor in the Vietnamese war in 1965. Earlier, around 1945, the United States was processing one-half of the world's fossil fuel. But by 1965 other nations had developed energy production, especially in the communist world (the Soviet Union and China), so that the United States was processing only one-third. Thus the American portion of world power had gone down. In 1945, the boundary of diplomatic and military influence of the United States was the edge of the Asian mainland; but this was no longer realistic in 1965. If we had understood that boundaries of influence must contract when a nation's relative energies do, the United States might have realized that its effective boundary of influence was further to the east.

WAR

The boundaries between competing energy systems may be stable if the size of each territory is in proportion to its relative productivity. When energy is rising, there is more tendency to grow and to expand influence on surrounding units. There may be a tendency toward more war when energy is large and rising—certainly there is a tendency toward more destructive war. When energy decreases, there may be less energy available for warfare and less tendency to try to spread influence.

If one area increases its true power over another but boundaries remain the same, there may be an unstable situation that could lead to war. If a nation thinks that it has the power to invade beyond its real power to dominate (as was true of Nazi Germany), or if it thinks it can defend a boundary where it no longer can, then some upset may follow. There may be economic upheaval, military action, revolution, or a combination of these. No one is sure, however, about the relationship between energy, territory, and war.

In general, the outcome of war is related to the energy available to the participants. If we could calculate in advance which side (if any) was favored in terms of energy, and if people understood the role of energy and were thoroughly convinced that these calculations would predict the outcome of war, then there might be boundary or other adjustments that recognized the realities of the energy situation without war.

UNITED NATIONS

The United Nations was set up so that each nation has one vote in the General Assembly; in the Security Council, each of the nations that were most powerful in 1945 has one vote. By the 1970s, as energy resources were being used up in some nations (like the United States), power was shifting. Many small nations with little energy were equal in voting power to the most powerful nations. Thus the outcome of a vote in the United Nations is not related to the distribution of true power in the world; consequently, it is not really possible to resolve a dispute by vote, since the votes do not represent power to enforce a settlement. It would perhaps be better to set up international organizations to reflect the reality of flows of energy and include some mechanism for adapting to changing energy patterns.

SUMMARY

In this chapter we have examined the uneven distribution of sources and flows of energy in the world. Because of the way energy has grown in human history, the distribution of excess population does not coincide with effective production of energy resources or with potential energy resources. Shifts in national power and dominance are related to the depletion of fossil fuels. Economic vitality depends on a nation's internal energy, since this determines a favorable or unfavorable balance of payments and the ability or lack of ability to obtain goods from abroad. Devastation of war is worst when energies are high. Wars are most frequent where energy is expanding or where there are changing relationships of energy. Present international organizations are not set up in accordance with the actual distribution of power, and thus there are no means of settling disputes that can respond realistically to the energy situation. If we understand how the distribution and use of energy control famine, war, and the survival of populations, we will have a better chance of avoiding the disasters that result from ignoring the realities of energy.

FOOTNOTES

1   From K. Hubbard, "Energy Resources of the Earth," *Scientific American,* vol. 225, 1971.

Chapter 13

Energy and the Individual

In simpler times the "energy life" of an individual consisted mainly of direct personal actions to draw food, clothing, shelter, and heat from the environment. The energy inputs to the individual came from long pathways of work by the natural ecosystems. See Figure 13-1. Today, in the high-energy urban world, the individual is less directly connected to natural energy flows. Instead, much energy comes from the human economy. As is shown in Figure 13-2, the energy flows to a modern individual are much larger and more roundabout: they come through many steps. Many of the inputs from purchases and public services are high-quality items, far from flows of raw fuel. We are not conscious of our large energy base. Because there are so many inputs and possible inputs to our daily existence, our world is more complex than that of our ancestors.

Figure 13-3 is a more specific diagram for an individual. It shows the energy flows and interactions for a person who teaches and keeps house. Each person is, of course, different; but patterns are similar from person to person.

The individual does not often think of his needs in terms of energy. He sees so many alternatives that he thinks himself free to make whatever he wants of his life. Indeed, in recent times, levels of energy have been so high and have grown so fast that individuals were seldom required to consider their

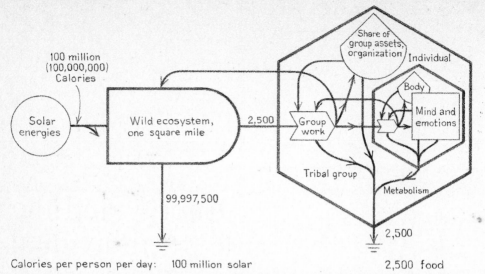

Calories per person per day: 100 million solar   2,500 food

Figure 13-1 Energy basis of an individual in a primitive tribal group sustained by hunting and gathering from nature. Numbers are heat equivalents per person per day supported by 1 square mile of the ecosystem.

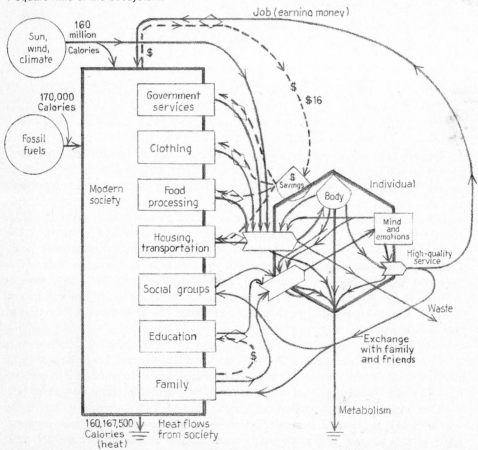

Figure 13-2 Energy flows to an individual in a modern society. Numbers are heat equivalents per person per day.

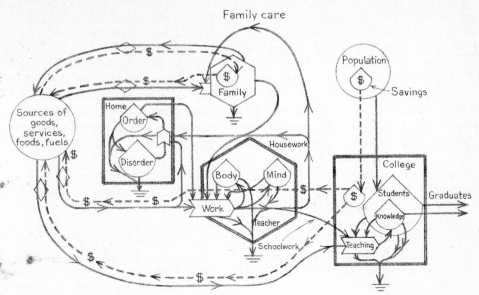

Figure 13-3 An individual energy diagram for a person who teaches and keeps house.

energy basis. For some, there has been no sense of duty to contribute work. There has been a sense of progress and change—so much change in some cases that the response to it has been described as "future shock." Individuals have lost the sense of their own worth to the world because there has been so much excess energy, so much variety, so much change, and so little required of them. Individuals have had difficulty recognizing their own place in the system, feeling themselves a part of it, or feeling valuable. Wars have become terrible for the individual and seem unrelated to individual morality. There has, of course, been an emphasis on individual rights; but the ideal of equal distribution was in conflict with a system characterized by unequal distribution. Some individuals have been part of the large flows of money and energy, but many have not.

The educational system has taught us more and more about bits and pieces, but less and less about overall meaning. Academic subjects have become fragmented and specialized. Commonsense ideas about the overall significance of energy are rare. The money system, with its depressions and inflations, seems capricious and mysterious. The individual has often been given much knowledge but little understanding. Old ideas of religion and duty seem irrelevant to ideals of personal freedom and self-fulfillment. Attempts to be individual sometimes leave the human being, basically a social creature, drifting and isolated.

As we have seen, a system whose individual members are not fully feeding back in exchange for energy received may not maximize its effective energy flows. When we recognize that there are energy laws controlling our patterns, we may well look again at our system as a whole to see how it should work and how individuals can make valuable contributions to it. Diagrams of our energy basis may show individuals how to help the system survive and give their own lives meaning in the process.

Figure 13-1 shows the upgrading of energy from solar energy to the high-quality elements of human life. Wild ecosystems, containing about one person per square mile received 100 million Calories of low-quality solar energy. This was converted into 2,500 Calories of higher-quality energy in the form of fruits, roots, and game consumed by human beings. One human in a hunting and gathering system required 2,500 Calories of concentrated food energy, or 100 million Calories of original low-grade solar energy, per day. Within a human system, this energy is again upgraded in three interacting pathways. One supports group activities, such as organization, teaching, and village management. The second is the operation of each individual's body. The third is the operation of the human mind and emotions. These are energy flows of very high quality that we call *information handling.* The individual feeds back this high-quality energy, as Figure 13-1 shows. As was mentioned before, the individual has a symbiotic relationship with the system of which he is a part if the energy effect of his special work is as great as the drain of energy produced by his energy requirements.

Figure 13-2 shows energy inflows to modern cultures, which involve fossil fuels and the many complexities that accompany the harnessing of high-quality energy to interact with solar energy. Money is involved in part of the energy flow, but not all of it comes from purchases. An individual usually obtains money by holding a job or being a member of a jobholder's family.

The energy flows in Figures 13-1 and 13-2 are given in heat equivalents, so that inflowing energy equals outflowing energy. To allow a similar comparison of energy flows, Figure 13-4 shows the same systems with energy flows expressed in fossil-fuel equivalents.

ENERGY PER PERSON

The heat budget of the food-consumption processes within a person is about 2,500 Calories a day. The energy base for all the special necessities is much larger, since it requires many materials and processes that are highly concentrated and upgraded. In primitive societies, the average energy flow per person per day in fossil-fuel equivalents was about 50,000 Calories. In the modern high-energy culture of the United States, the energy flow per person per day is about 250,000 Calories (see Figure 13-4). The high-quality education and complex activities of the modern world are made possible by these large energies.

The feedback that individuals must supply to justify such expenditures of energy is very large and requires experience and training. Many years are required before the energy effect of an individual equals the cost of that individual. Many people never make meaningful contributions. Young people find it difficult to get jobs where their work puts enough energy into the society

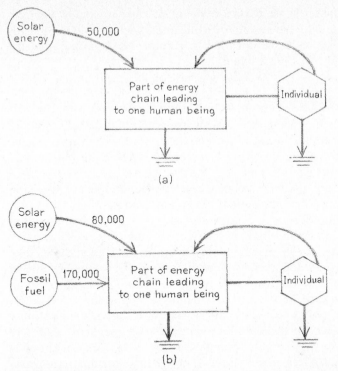

Figure 13-4 Energy flows supporting an individual expressed in Calories of fossil-fuel equivalents. (*a*) Primitive culture (see Figure 13-1). (*b*) Modern culture (see Figure 13-2). In these diagrams, solar energy has been converted to fossil-fuel equivalents by dividing by 2,000 (see Table 6-1).

to pay for the high energy cost of their education. The simpler jobs and roles are occupied mainly by machines.

DISTRIBUTION

Early in the industrial revolution, the discovery and use of new energy produced assets and concentrations of money in expanding new companies. The role of fossil-fuel energy in all this was not entirely understood. Those who had worked to start companies and make the new systems go believed, partially correctly, that their special effort had generated all this wealth. They believed that the concentration of money and power in their hands was a reward for their work. They did not realize the value of fossil-fuel energy.

As industries became more mechanized, people performing small parts of processes in factories could not see the energy value of their work. Their goals became money and leisure. Because their work did not seem to be meaningful, they often spent their leisure time doing creative things such as sewing or carpentry. When people lost the sense of the value of using their own energies for work, they often lost the sense of their own value. They could no longer see themselves as part of a valuable system of using energy for productivity. This feeling of worthlessness has sometimes led to mental illness, crime, and the attitude that "the world owes me a living."

The distribution of the new wealth was not equal, even among people working in the new industries. This has added to the turmoil throughout the world over uneven distribution. Labor unions and progressive taxation developed in the United States because of a widespread feeling that distribution should be more even. Some countries, such as the communist countries, have tried new systems, without much private enterprise, in which the government regulates more of the production and distribution. These countries have had less variety, flexibility, and individual choice and have generally been slower in developing their economies in international economic competition.

By the middle of the twentieth century, the gap between rich and poor in the United States, though smaller than in many other countries, was nevertheless great. Energy and assets had become much larger in the United States than anywhere else in the world, and a more equal distribution of wealth became an ideal. A minimum standard of living was proposed. More equality of food, shelter, clothing, health care, and access to educational opportunities became a main concern of public policy. Equality and justice were advocated, not only because of value they had for energy-effectiveness (since a greater equality in the distribution of energy may increase people's ability to do work), but as ends in themselves. Equality *itself* became the ideal, rather than equality as a *means* of making life more energy-effective. Machines took more and more jobs away from people, and people without a good education were unable to find jobs that paid much. Minimum-wage laws were established, but they made it difficult for the very young, the old, and the untrained to find jobs at all, since many of these people do not have training and ability worth the minimum wage.

A great experiment was undertaken: taxes were levied on those with good incomes, not only for necessary government functions, but also to pay the poor. In other words, money representing some of the energies of the society was routed to people without direct feedback from them to help the economy work more. The purpose was to help the poor join the mainstream of work. But there was a basic problem: there were not enough meaningful roles, because available energy was mostly supporting large-scale systems and machines. People receiving payments were not always given work and were left to their own devices. The various things they chose to do did create diversity and sometimes led to new approaches which were ultimately energy-effective.

FREE WILL AND ENERGY LAWS

We have already seen how the parts of a system help the system to survive by feeding back high-quality work in exchange for energies used. A human system maximizes power if its individual members, in exchange for their own support, contribute back to some other people, to some part of the machine system, or to some part of the natural system. Most of us do this by getting a job, and the economic system arranges the matching of our services with the input from the system to us. In addition, our family structure and our system of coupling, love, and affection arrange cooperative actions. On a small scale we readily feel the effectiveness of our small groups. But we have more difficulty relating to the larger systems of energy flow, since our emotions are not readily attached to nonhuman, large systems. Education is required to help us see the workings of the larger institutions. (See Figure 13-2.)

No system can understand itself, since more intelligence is required to understand than to be. Individuals try millions of activities without much consciousness about how these might affect other parts of the system. This is all right, because an activity which is energy-effective will generate a feedback reward that causes it to be repeated or to grow. Individual freedom is a form of disorder that generates order when it interacts with unused energy. There is apparently a balance of order and disorder (see Chapter 3, Figure 3-1), and individual initiative and coordinated work are a part of it. When energy flows are large, then there is more energy for individual freedom.

Another tendency in individual behavior is initiative—the urge to be creative and different. This too has important effects on energy. Keeping the system of humanity adapted to energy conditions requires a continuous creation of choices and alternatives. Many of these turn out to yield no net energy and are not reinforced. A few are energy-effective—they succeed in getting more energy for the system and the individual—and are copied by others. However, alternatives tried and discarded are just as important as the successful ones: they are necessary in finding the successful ones. Creativity and individual initiative are mechanisms for choosing—ways of adapting to energy conditions. The individual has free will and contributes by exercising it, whether or not his activity is chosen or discarded by others. The energy laws ultimately control the role of humanity, but individual freedom helps keep humanity adapted to its energy flows.

THE WORK ETHIC AND PROGRESS

In cultures with maximum growth during times of expanding energy, individuals were trained and rewarded for hard work. Dedication to individual work is sometimes called the *work ethic*. Many schools and religions teach it.

In other cultures, at other times a high level of individual work was required for survival without growth because there was little in the way of special energy sources like fossil fuel. In a culture operating on the renewable energy of nature without any large reserves, little growth was possible; the new simply replaced the old. To contribute to the survival of that system, the individual needed a sort of work ethic, but it did not include expansion and progress. Work is necessary for survival in both growth and no-growth cultures. What distinguishes the work ethic is its emphasis on growth and progress as good. Since growth and progress may not be possible except during periods of energy expansion, individuals may now have to reconsider their attitudes. Work and growth may have to be considered separately. In periods of steady energy *success* may come from "contribution" and may not include expansion of activity, wealth, or power.

EMOTION AND ENERGY

A human being carries many inherited aspects of emotion and instinctive response. As we mature, we learn to fit these inherited tendencies with our learned tendencies to make a functional whole. Some of our emotions, such as sex, fear, love, hate, and aggression, consume much of our time. Thinking about the emotions, organizing them, and dealing with them represent much of the flow of energy that supports the individual. Whether an emotion is suppressed, emphasized, or converted to an activity, it must be dealt with. Emotions may be used differently within a culture or by different cultures. (For example, emotions between the sexes may be used in reproduction, in improving social relationships, and in advertising.) But since human survival depends on many activities, a successful culture will somehow ensure that no one activity (such as sleep, play, or sex) takes up too much energy.

Aggressiveness has sometimes been regarded as creating in human beings an innate predisposition for warfare. We suggested in Chapter 12 that war is more a function of the large-scale energy trends than of individual reactions. Aggressiveness, like other emotions, is generally put under control by individuals, although it may be released by group actions in ways that make it useful. Each successful culture finds ways to adjust human emotional tendencies so that they contribute to energy rather than drain it. Individuals unconnected with any social organization may have great difficulty in maintaining their balance and functioning well, since the human being is ordinarily a social organism. We usually operate our energy flows so that *group* responses give us positive feedback; this makes us feel our roles are valuable.

AESTHETICS

Beauty and other aesthetic values are important to the individual. We often think of these as impossible to measure, because they are so individual and are often, by agreement, left up to individual choice and preference. Sometimes, however, we do try to evaluate aesthetic preferences by such means as polls and questionnaires.

Beauty and other aesthetic values do involve energy. First, energy is needed to develop the *ability* to create aesthetic qualities. For example, much energy goes to support the institutions which supply the long education and training necessary for creating and maintaining art. Much natural energy, for another example, goes into a forest which serves to renew the spirit of those who visit it.

Second, energy is involved in conditioning (that is, educating or training) people who are consumers of art. For example, some paintings are widely regarded as valuable because millions of people studied them in "art appreciation" courses—an effect that obviously has entailed a great expenditure of energy.

The aesthetic characteristics of a culture have a role in its energy flows. Some attributes of form and function develop through education and tradition, and in turn cause energies to flow in their support. Green grassy plots (our lawns) are a very energy-expensive aesthetic preference of our culture. Because pioneers in forest land developed grassy green plots of new agriculture, grass may have become a symbol of survival.

When energy is rich, there may be more time and resources for aesthetic creativity, and thus the arts may flourish more in these times.

LEVELS OF ORGANIZATION

Human systems are composed of many levels of groups into which each individual fits. There is the main governmental sequence: family, town, county, state, nation. There are groups having to do with organized work, religious activity, and recreation and social exchange. The individual allocates some time and effort to developing exchange with others through these groups. But an individual's social capacity is limited to a relatively few meaningful, continuously renewed connections. The exchanges of information and action in these pathways are high-quality energy interactions which generate self-organizing reward loops that support the individual. An individual is healthy if there are a reasonable number of these interactions and choices in his life.

At times, when the larger system is well organized, individuals may shift their initiative and allegiance more to smaller groups, causing more choice and variety to develop. In such times, individuals may work for human goals directly, as in social work. On the other hand, in times of great need on the large scale—times of national crisis—individuals may turn their efforts in unison to the goals of large organizations so that the system can survive.

HUMAN ROLE IN THE CHAIN OF ENERGY QUALITY

We have already discussed chains of energy—chains in which each stage generates higher-quality energy than the preceding stage but uses up some of the available energy in the process. An energy chain from sunlight to electricity was shown in Figure 2-4. Figure 6-1 also showed a chain, indicating how energy is concentrated in space as its quality is increased. Energy chains in ecosystems (food chains) were shown in Chapter 7, and energy chains in earth processes were shown in Chapter 8. Figure 13-5 shows the general pattern of these chains. Figure 13-5a is a simplified chain of energy users. Flows of energy decrease going from left to right, as shown in Figure 13-5b; but the energy flow increases in quality from left to right, as shown in Figure 13-5c.

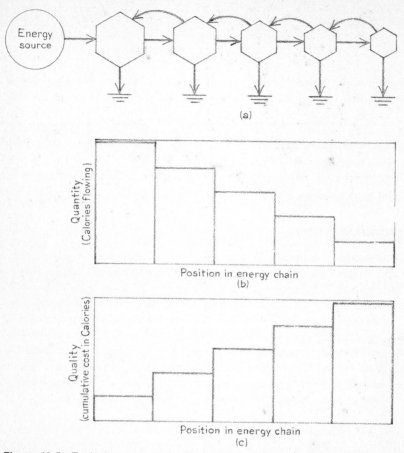

Figure 13-5 Typical energy chain: the quality of energy goes up as the quantity of energy goes down. (*a*) The energy chain. (*b*) Quantity of energy flow per unit of time (heat equivalents). (*c*) Quality of energy flowing. (The measure of quality is Calories of fossil-fuel equivalents per Calorie of heat equivalents; see Chapter 6 for the explanation.)

Similar chains exist within human occupational roles. Different human beings occupy different positions along the chain. Although human activity in general is high-quality, some activities are farther to the right than others on the scale of energy cost and energy quality. Jobs which require unskilled hand labor are on the left; jobs which require long periods of education are on the right. The jobs on the right tend to pay higher salaries, but there are fewer of them. Large numbers of people are needed for jobs on the left; few people are needed for jobs on the right. However, since the industrial revolution there has been a tendency for machines to replace people in jobs on the left. The industrial revolution has also increased the total energy flow, lengthening the chain of energy quality (see Figure 13-6), and this has increased the number of jobs on the right.

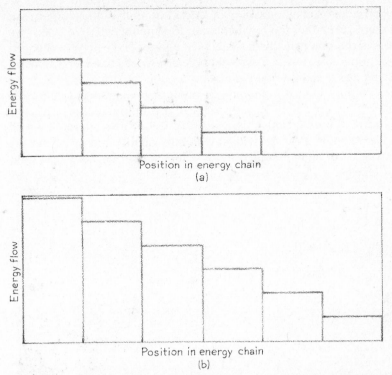

Figure 13-6 Comparison of the energy-distribution graphs for (*a*) a low-energy system and (*b*) a higher-energy system. With more energy, some additional units of very high quality can be supported. (Units are Calories of heat equivalents flowing per unit of time.)

SUMMARY

In this chapter we have examined the individual as a part of large and small systems of energy flows. So long as the individual finds opportunities to contribute, he receives in return the means of support. Because there is a great variety of places for the individual to plug into the system, there is a great degree of choice left to him. Individual initiative helps ensure that the energy contribution of individuals is maximized. By the creative actions of individual initiative and intelligence, energy systems can remain adapted to reality. The survival of the supporting systems and thus of individuals depends on the action of the individuals to fit themselves into these systems somewhere. How to do this best is partly left to the individual, who must find a use for his background and individual resources. From the point of view of energy the individual is important, not per se, but because of his opportunity to make a contribution. An individual may be useful by joining in tasks that need group effort, by doing separately specialized jobs, or by creating ideas that might be useful later. The images and ideas that people use to guide their work need not be consciously based on maximization of energy if that is their effect. An individual need not feel either too regimented or too unconnected, for there are all kinds of roles in the small and large systems of energy that can be exchanged for support. We apparently have an inherent or learned need to be a meaningful part of something larger than the individual.

The energy supporting an individual in modern industrialized culture is about five times that supporting an individual in a primitive culture. Energy support in modern cultures includes the energy of fossil fuels interacting with solar energy. The support of an individual with high-quality energy causes his or her activities to be of higher quality, more complex, and farther to the right in the chain of energy.

Energy Crisis and Beyond

In Part One we discussed flows of energy and the laws by which energy is used, maintains order, causes cycles, and determines what pattern survive. On our planet, flows of energy turn the atmosphere, oceans, landscapes, and human ecological and economic system.

In Part Two we examined the energy basis for human life in preindustrial times and during the industrial revolution, including special processing of energy by human beings, international exchanges, and flows of energy to the individual.

In Part Three we consider events and changes that may be considered to have started with Earth Day in 1967, the energy crisis in 1973, and the surge of inflation and economic stress in 1974. We will look at some hopeful trends, which may point toward a steady state. Much of what is happening can be understood in terms of net energy, a concept we have already discussed.

As individuals and institutions feel the impact of inflation stemming from a shortage of energy, people are becoming more receptive to new ways of looking at the energy policies of local, state, and federal government and international organizations. To clarify the issues, we must understand how changes in energy are acting on our society, why some recommendations contain errors, what energy conditions may be in the future, what kind of life lies ahead, and how a smooth transition can be made.

Part Three has three chapters: Chapter 14 analyzes the changes taking place as we shift away from a growth economy; Chapter 15 describes the stable pattern we may expect if energy flows become steady; Chapter 16 summarizes our outlook for the hopeful future.

Chapter 14

Energy Crisis and Inflation

In this chapter we study the causes of the energy crisis and our recent inflation, discuss how energy principles explain and predict trends in the economy, examine errors in recommendations concerning energy, and suggest ways of making the best transition. Let us begin with the situation that existed at the time the oil-producing nations raised the price of oil in 1973.

NATURE OF THE ENERGY CRISIS

By 1970, urban growth throughout much of the world was extraordinarily steep. Compared with anything that had happened on earth before, the explosion of humanity was staggering. There had been many warnings that resources and energy were insufficient to continue the accelerated growth of populations, machines, and urban development. Perhaps the first danger to be noticed was the pollution of the biosphere; there was a surge of interest in protecting the environment. However, the trends of two centuries remained in our cultural memory, and many aspects of the life-style that went with the expanding economy were not readily put aside. It was customary in many courses of economics to make a joke of Malthus' warnings of two centuries earlier. Young people were questioning some of the excesses of the high-energy world, its wars, its absurdities of advertising, its unrenewable cities, and its growth ethic; but most people all over the world were still expecting new, easy sources of energy to be developed.

In the early 1970s the energy cost of getting energy began to increase so much in many countries that the energy available for other activities began to be scarce. As soon as some shortages of energy developed, nations with rich deposits of fossil fuel were able to raise their price: the price of oil went up by a factor of four in the fall of 1973. Nations which had to buy their energy found that the net energy available for their own assets and growth was sharply reduced. Growth slowed or stopped in all but the energy-rich nations. Major changes in economies, international relationships, and the lives of individuals began to take place.

At the same time that the price of oil went up suddenly, the oil-exporting nations also decided to reduce the output of oil. This reduction created a shortage in the oil-consuming nations. It was difficult to get gasoline for automobiles and enough heating oil to keep homes at normal temperatures during the winter.

This energy crisis was regarded as a temporary, artificial shortage of energy, caused by monopoly. The response of those in energy-consuming nations was to make temporary adjustments to restrict consumption, consider rationing, set up federal organizations for managing energy, and increase expenditures for new energy options. Most citizens did not understand the connection between the leveling off of the economy and the energy situation. Many parts of the economy kept trying to grow, even though it was becoming increasingly clear that energy resources for growth were not readily available. Finally, though, people are beginning to learn about their energy basis. They are ready to ask: What kind of world may we expect in the future, with less and less energy available?

ERRORS IN THINKING ABOUT ENERGY

The public and its leaders in the United States and many other nations have tended not to believe that there is a shortage of energy. They have supposed it to be a shortage of refining, drilling, coal mining, and other activities needed to tap what they believe are unlimited energy resources. This confusion has been fostered by several errors in thinking.

1 Economists have taught that there is no such thing as a shortage that raising prices will not cure. But the economists have not been educated in energetics and therefore have not understood the second law of energy and the fact that energy is not reused. Economic advisors often recommend policies that are meant to encourage high prices, believing that high prices will end shortages of energy and not realizing that more energy in the consumer sector is directed to getting energy regardless.

2 Although most people are taught about energy in their schooling, they have not been taught that forms of energy differ greatly in their ability to contribute to useful work. Many scientists and architects have pointed to the large amount of energy Calories reaching the earth each day and have suggested that some technology could harness it. What they do not realize is that most of this energy would have to go to pay for its own concentration. Along with a blind faith in human technology, many people have a blind faith in the eventual availability of new energy.

3   Many people have been influenced by estimates of the oil and coals yet to be discovered or mined in the United States. Very large figures, such as "resources for a hundred years," have been given out. Some of these are based on the calculations of the *chance* of finding new deposits based on previous drilling. Much of the land left to be explored, however, is less likely to have deposits because of its geological history. Moreover, most of these figures consider that coal deep in the earth and oil far out to sea is as good as that nearer the surface or shore. Thus the high cost of the special kinds of technology needed to tap those reserves is not included in the figures. Some of the reserves do not, in fact, represent a gain in net energy.

4   People who have seen in their own lifetimes the development of automobiles, airplanes, radio, television, control of most communicable disease, and a space program which put men on the moon, often draw the conclusion that there are no limits to human ingenuity. Most people believe that technology can do anything. Engineers and many technologocal scientists have tried to caution us, but faith among people who know no science dies hard. Journalists and politicians have more economics than energetics in their background and often simply have not believed those who try to explain the limitations placed on technology by a shortage of energy. Technology in the twentieth century has been mainly the ingenious application of fossil-fuel energy to help various processes. An energy shortage means that technology cannot be applied so much. There have been many specialists seeking more money for research to develop new approaches, using technology to go after various types of energy. The public has not questioned this; those who showed the errors in their calculations have been regarded as defeatists.

5   When higher oil prices brought more energy in, many people assumed that their disbelief in the energy shortage had been justified. The huge inflation caused by the high price of fuel was regarded by many as something apart from the energy shortage. The more inflation has hurt the economy, the more they have looked to economics instead of to the basic cause, the increasing energy cost of finding energy.

6   When fuel prices jumped, anyone who had oil in a tank or in a tanker could sell it for much more than he had paid for it. Those in the fuel-processing business made enormous windfall profits. The public regarded these profits as evidence that the whole energy shortage was contrived—a trick to make profits at the expense of the public.

7   Many people believe that nuclear energy will supply what is needed. Atom bomb explosions during and after World War II demonstrated the enormous energy of nuclear reactions. An army of scientific optimists set to work developing atomic energy; by 1975, 100 big nuclear generating stations were established. This effort used much of the nation's brain power, capital,

and options for energy. But the very large energy cost of developing atomic energy in power plants was not the concern of the optimistic physicists, nuclear engineers, and economists. Their equations did not take account of the hidden subsidies to nuclear energy from fossil fuels. The public does not have the overview that would allow it to see that nuclear energy is not very rich when controlled. Because atom bombs are so powerful and so hot, much of the energy produced must be used to control the processes. This does not leave much for production of energy useful to man.

8 The idea that growth is necessary for prosperity is ingrained in industrial cultures. Their leaders seldom if ever allow themselves to mention the possibility of stopping growth or of a decline in assets, energy flow, and the economy as a whole. To admit the reality of the energy shortage has seemed to be an admission of nongrowth, extinction, and doom. The possibilities of good life without expansion have not generally been considered.

9 Because of the complexity of the long webs of energy flows, most people have judged the higher costs of energy in their lives as small, since the percent of energy which they paid for directly was small. For example, suppose that a person thinks of his fuel bill as only 10 percent of his personal budget and of the cost of crude oil as only a small percent of that. What he fails to consider is that everything else he uses also comes from fuel and will therefore eventually cost more. He does not realize that ultimately everything is based on energy and will jump in price. The further an item is from oil in the energy chain, the longer it takes for the price to rise. But the longer the chain, the more parts come together, each bringing an increase in cost. The jump in cost finally gets back to the energy supplier, who must raise the price again. The fuel supplier will raise his price once when the cost of fuel supplied to him goes up, and again when the effect of his own price increase travels through the economy and back to him in all the things he buys. People were misled when the first price increases were small.

10 Many economists have stated publicly that the price increase due to oil was a temporary effect. They have implied that adjustment of all earnings would take place without any ultimate effect on people's purchasing power. They do not realize that the relative cost of oil to other things has been permanently increased by the greater feedback of effort necessary to get oil. The oil supplier who has inferior deposits must have the high price to pay for all the inputs required. Suppliers drawing from inferior oil deposits do not get rich and ultimately go out of business as their energy costs get higher than their energy deliveries. An oil supplier who still has rich deposits (close to the surface) can increase his price, make a profit, and grow.

SURGE OF INFLATION

The increase in fuel costs in 1973 was a sharp reversal of a long-term trend: energy costs had been dropping. The inflation rate, which had been about 5 percent a year, jumped to 10 percent or more. Some nations which depended completely on Arab oil developed inflation rates of up to 25 percent. One of the first reactions of governments was to try to get more energy and more productivity by starting national projects. These caused more money to circulate, making the ratio of money to energy even worse.

The increase in the cost of foreign oil has meant that more energy value in sales must be sent abroad to exchange for oil obtained there. The net energy to the United States was, therefore, less. With less net energy for the American economy, the money circulating for the work done here was less. The dollar lost buying power; that is, it was inflated. The higher prices of fuel have made drilling and pumping of domestic oils profitable for deposits deeper in the ground and further offshore. There has been a new effort to get these deposits into production. Since these are deeper and further offshore, this means putting more of our energy back into getting the oil—building enormous steel rigs and pipelines and repairing storm damage. The net energy of the oil is less than before, while the amount of money circulating is the same or larger. Inflation has continued, driven by the increasing energy cost of getting energy. All net energy that was used before for the growth of the American economy is now going abroad, to swell the growth of the oil-rich nations, or into the oil companies, to support the huge expensive offshore installations and the equipment for deep mining.

SHIFT IN PROPORTIONS OF ASSETS

Our fuel reserves being used are becoming less concentrated and less accessible—deeper in the earth, further offshore, or further away in the Arctic. More and more of the energy obtained goes directly and indirectly into the energy industry for the work of processing energy. Less and less comes through to the main economy. The situation is diagramed in Figure 14-1. In Figure 14-1*a*, relatively few assets and relatively little money are tied up in the energy industry; the main economy develops large assets. The rich fossil fuels support the assets of the main economy and indirectly feed back some assets to amplify the solar sources such as agriculture and forestry (see the path labeled F).

In contrast, Figure 14-1*b* shows the situation when the fuel reserves are dilute and far afield. There is an enlarged energy industry (oil rigs and nuclear plants), and more of the economy is tied up in it. Less energy comes through to the main economy, and the assets stored for individuals are much less. Eventually, the energy industry may be operating with a net loss. This loss could be made up by the feedback assets of the main economy, running on solar energy from farms and forests. Solar energy is a steady, renewable flow. In 1974, there was an evident shift to a higher proportion of assets in the energy industry, with the main activity and the most available money in the oil companies.

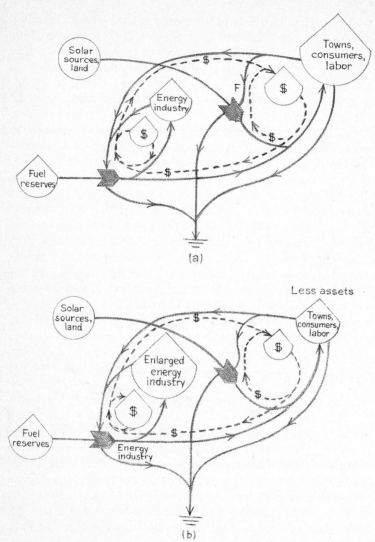

(a)

(b)

Figure 14-1 Effect of concentration of energy on the proportion of assets in the energy industry and main economy. (a) Rich energy, with much net energy going to the main economy. (b) Poor concentrations of energy, with more assets required in energy processing and allowing little net energy to go to the main economy.

FALLACY OF "INDEPENDENCE NOW"

One of the immediate reactions to the oil shortage has been a national effort to achieve independence of foreign sources as soon as possible. The implication of those advocating this policy is that the only obstacle is a shortage of industrial facilities for processing fuels. There is a widespread belief that there is no real shortage of energy—only a shortage of refineries, oil wells, coal mines, etc. Those pushing independence believe that higher prices will give the energy industries enough funds to develop all necessary facilities and process all necessary energy.

Such efforts made by nations which are short of energy will have the effect of using up what energy they have even more quickly than before. By becoming more independent *now*, these nations would certainly run out of resources even sooner and become even more dependent on others. The policy of independence actually works backwards: those who pursue it will be independent now but more dependent later. Since the foreign oils are richer, they will tend to be used first, in spite of national policies.

MILITARY INTERVENTION FOR ENERGY

Some people have reacted to the energy crisis by advocating military invasion of other nations to control the price of energy. However, the military strength of nations that are primarily energy consumers is not enough to guarantee that any such application of force would be successful, especially since there are other blocs of power in the world. The existence of nuclear weapons and the danger of military intervention which goes beyond a nation's actual power makes this solution of questionable value. As the net energy of a nation becomes less, the energy available for external affairs also becomes less. Nations with little energy are in no position to fight or even to threaten when the opposition is rich in energy. For them, safety lies in conserving their energies for local defense until others have also reduced their energies. For the good of all nations, an even distribution of energy is best: that is, more stable.

INCORRECT ASSIGNMENT OF ENERGY COSTS

People estimating energy requirements in order to find which processes are energy-expensive have made detailed calculations of energy requirements for various processes by estimating the fuels required. But the other goods and services required were not regarded as energy and so were not included. By ignoring required inputs, these people have often missed the very high energy requirements that went into goods and services processed earlier. Some complex human activities—like education—seem to require little energy; but if one traces all the complex inputs back, one finds that the energy involved is actually very large.

Figure 14-2 shows two industries each of whose output is required by the other; each is justified by supplying the other. One might casually conclude that the first industry requires more energy than the second; in fact, however, both have the same energy requirement. Both need everything that the input energies supply.

Another error occurs when all energy is considered equal; for in fact some high-quality energy requires more lower-quality energy to generate it. People who count electricity the same as coal are making an error, since electricity requires 3.6 times as much energy for its development as coal.

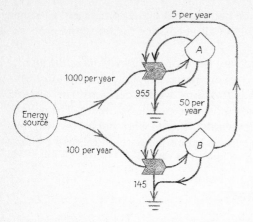

Figure 14-2 Two industries, *A* and *B*, which receive each other's products as necessary inputs. Although energy inflows seem different if one examines only the low-quality energy inflows, an examination of the whole system shows the energy needs of both to be the same. The direct fuel inflow is misleading.

ELECTRICITY AND POWER PLANTS

Although many people think of their source of energy as electricity, this form of high-quality energy is not an ultimate source. As we showed in Chapter 11, 3.6 units of fossil fuel are required to generate one unit of electricity. An increase in the price of fossil fuels has a very sharp effect on the price of electricity. Not so long ago, there were advertising campaigns to sell electricity for many purposes that did not really require such high-grade energy. For example, the "all-electric house" uses electricity for heating as well as for appliances. But it is much more efficient to use oil to heat the house than to use the oil to make electricity to heat the house. This is using high-grade energy for a lower-grade purpose that could be accomplished with lower-quality energy: it is a waste. When fuel prices go up, people living in all-electric housing are in great difficulty; many find that they must cut consumption or shift to lower-cost heating.

Public and private utility corporations have been building power plants to meet the "demand" for power. They were a growth-stimulating industry, since they developed the means for growth which attracted newer industries and population. But when the utilities tried to increase their prices, consumers were pinched and reduced their demand. Consequently, the utilities had no further money for expansion; and since capital was generally unavailable, they could not borrow it. In other words, with no net energy there was no money or energy for further expansion. Some of the expansion that had been planned was nuclear and was supposed to substitute for declining fossil-fuel supplies. Without net energy in the existing plants, there was no means of making any further transformation to nuclear energy. In 1974, only 5 percent of our electricity came from nuclear energy. The net energy contribution from nuclear plants was less than the decline of fossil-fuel net energy.

SHORTAGES OF RAW MATERIALS

In this book we have considered raw materials such as iron ore, copper, and aluminum ore as energy sources, much like the fuels. Their contribution to the production of goods and services has an energy component (see Figure 11-15). When these materials are rich and near the surface, they take little other energy to process: their energy contribution is then large. However, we have been mining deeper and deeper and using poorer and poorer ores, and to do this we have had to increase the amounts of fuels that we put into mining, concentrating, and processing these materials. These energy costs went unnoticed so long as the fuels were in excess. Now, with sources of raw materials harder to reach, and without cheap fuel energy to process them, a shortage has developed. When fuel costs rise, the prices of these raw materials increase sharply, since such materials as steel, copper, and concrete are among the largest energy users.

Some people see this situation as a shortage of raw materials. But since inaccessible and dilute ores represent poor-quality energy, the problem is really an energy shortage. Our problem is that our remaining sources of both fuels and raw materials are low in net energy. During the energy decline, as total activity decreases—that is, as use declines—surpluses may appear, causing prices to fall; but this is a temporary effect. Ultimately, the increasing energy cost of getting energy takes a larger part of the economy.

INTEREST RATES AND LOANS

In Chapter 9 we showed that an absence of net energy and growth also eliminates money for loans. A loan will not generate any new productivity if there is no net energy to be tapped. If there is no production, the loan cannot be repaid. And if a loan cannot be paid back with interest, it will not be loaned. If money is held, it loses value because of inflation. A loan must have an interest rate equal to the rate of inflation rate (to keep its value) plus some additional interest (to make money). If there is little net energy being generated, loans cannot be made without bankruptcy (that is, they are never paid back). If there is also inflation, then the interest must be even higher, and the chance of the loan's being paid back is even less. Growth is usually attempted by borrowing to start new activity. A scarcity of energy along with inflation thus stops growth. An example at the time of this writing is the threat of bankruptcy in New York.

Money that banks loan out comes from their depositors, to whom they pay interest of 4 or 5 percent. The banks then loan it out at higher rates of interest. If the inflation rate is much higher than the interest paid by banks, banks will not get the money. Instead, people will buy goods and services that depreciate at a rate lower than the rate of inflation.

Money for loans also comes from central banks and international sources, but if a nation has an unfavorable balance of payments, more of its money goes abroad and is unavailable for loaning at home. Inflation and an unfavorable balance of payments because of energy shortages also stop loans and growth. This becomes easier to see when we compare money and energy in the diagrams (see Figures 4-2 and 10-4). Consider that the energy-production flow sends through less energy and the money inflow stays the same or increases.

STOCKS AND BONDS

Stocks and bonds are sold by corporations and governments to get money to start new projects, such as power plants. Stocks are shares in a corporation. Bonds are promises to pay at a later time with interest. Government bonds may be free of tax and thus may be worth several percentage points more a year than their rate of interest might indicate. Stocks and bonds are instruments of developing capital for expansion. People who held them during the growth era expected to get interest from them; they expected their money to earn money and took it for granted that this would always be so.

When net energy is small, there is little growth and thus little increase in money. There is no profit to those with money: their money depreciates or earns very little. There is, of course, little advantage in owning stocks and bonds if they are losing value. When this is the case, more people sell stocks and bonds than buy them, and the value of the stocks and bonds on the market goes down and down. Companies and governments trying to sell stocks and bonds cannot sell them and thus cannot get the capital for new projects. They cannot grow, because ultimately the energy for growth is not available.

DEVELOPMENT OF NEW ENERGY

The immediate, reflexive action of the national government to the energy crisis has been to start projects to get more energy—such as more regular (fission) nuclear plants, more energy-expensive nuclear processes like the breeder process and fusion, and more offshore drilling. However, the decline in the net energy in purchases of foreign fuel (caused by high oil prices abroad and the decline in net energy involved in developing our own, poorer deposits) has reduced capital by eliminating growth. The very large new projects cannot, therefore, be accomplished. There was some capital available from the profits of high-priced oil, but it was soon to be lost in buying steel and other materials as higher energy costs were translated into higher prices. Capital is being spent on getting energy from larger energy-mining structures, and is thus shifting away from consumers (Figure 14-1). With capital being used in primary drilling and mining, there is little available for processing energy further from the source. Utility companies, finding all their energy costs high and the demand for their product decreasing, have had to cancel the building of new plants.

INDIVIDUAL BUDGETS

Inflation began to lower the individual's energy budget as higher prices worked their way from energy on through the economy. At first, many people were not especially concerned, thinking that they could maintain their buying power by means of raises in salaries and more favorable new labor contracts, as they had always done during periods of expanding energy. However, we have found that companies are getting less net earnings—and county, state, and federal governments less money from taxes—than had been anticipated as growth slows down and stops. The simple truth is in the diagrams: less net energy means less of everything per person. Energy controls the economy.

INVESTMENT BY OIL-RICH NATIONS

A major question has been whether the oil-rich nations would invest their net accumulations of money back in the industrialized countries. If this happened, there would be continued growth in the Western nations based on the rich oil, except that the Arabs would be the managers. Some economists believe that this is the only way the Arabs can spend their money. In fact, there have been purchases by the Arabs, but there were many reasons why these are not large enough to allow continued growth of the industrial nations:

1 Foreign investors worry about confiscation. There is a joke that goes like this: Let's sell General Motors to the Arabs and then nationalize it. Outright confiscation might be unlikely in the United States, but regulation of profits and monetary policies might have the same effect.

2 If investment were made outside of the Arab nations, the net energy invested would develop the real assets of the nation where it was invested. This would go against feelings of self-interest and the nationalistic desire for control.

3 The Arabs could invest abroad in order to keep their money from depreciating as a result of world inflation and currency devaluations, if they could get repayment with reasonable interest. They would, of course, have to charge interest higher than the rate of inflation. But the Western countries, with their rising prices and energy shortages, could not borrow under these circumstances—or, if they did, they would go bankrupt. There have already been some bankruptcies in large, prominent firms that are beginning to cause managers to borrow with caution.

4 In more general terms, in the industrialized nations growth is stopping and net energy and profits are disappearing. This kind of economic climate makes the industrialized countries seem unlikely places for investment.

5 A better alternative is available to the oil-supplying nations. They can use their money to buy and bring home assets, including technology, equipment, personnel, raw materials, water, and even land of adjacent countries. This trend has developed. Citizens of the Arab nations are going abroad in great numbers for training in order to bring back know-how. There may be some similarity here to the rise of Rome, for Rome imported much of its knowledge from Greece.

6 Being concerned about the possibility of "boom and bust," Arab nations have used some of their wealth to start nuclear power plants. Their supplies of uranium, however, are not very secure, and the net energy of nuclear processes is not very certain, as we discussed in Chapter 11.

7   Always faced with the possibility of armed intervention to capture their oil, the oil-supplying nations have put a great deal of their wealth into military development. They have bought their armaments from the West. Their purchases have included the means of developing some nuclear capability. The rest of the world is now naturally concerned lest these areas of change and expansion should be so energized as to start an atomic war.

INFLATION AND UNEMPLOYMENT

The United States has deliberately maintained an inflation of several percent for the last few decades, since inflation was stimulating new net energy and thus stimulating new jobs. Creating new jobs at the cost of inflation was accepted by Americans. The same effect could have been arranged by taxing them and using the tax money for developments. The approach through private enterprise was to loan money in excess of taxes, causing inflation by adding new money faster than new energy developed. Deficit financing of this type was the same as a tax, since all the citizens lost buying power to help pay for the new enterprises. As long as the new enterprises developed new energy inputs, this kept energy flowing and maintained a valid, competitive economy.

When inflationary spending no longer brings in new energy, it generates no new jobs and is only a disadvantage for citizens. Each person is being taxed, but the money is put into nonyielding and futile activity.

Economists have feared unemployment most, remembering the serious depressions when money circulation stopped even though there was plenty of energy. They have been afraid that if money is not stimulated by loans and growth, massive unemployment will result.

However, the condition in the 1970s is different because the inflation is caused by failing energy. Since money loses value if it is held, everyone is being stimulated to spend it rather than hold it. Even in banks, money has lost buying power each year. Consequently, more money has been converted immediately to food, land, etc., and as a result money has continued to circulate well even though it is losing buying power. In general, energy for machines has been declining and many energy-rich industrial activities—including agriculture, housing, and fabrication of raw materials—have been decreasing their use of fuel and increasing their use of human (hand) labor.

However, a decline in energy means a decline in jobs in other categories. The change is like the difference between the graphs in Figure 13-6a and b. Unemployment has been prevalent among specialists and those working in high-energy parts of the system. The shortage of jobs has caused trained and skilled workers to displace less-skilled workers from their jobs. The number of jobs affected on the right side of the energy chain in Figure 13-6b is large, but the number of jobs affected on the left side of the chain is larger. There may be more jobs in agriculture as machines are used less; but this trend may not appear until machines are worn out.

ENVIRONMENTAL PROTECTION

The energy crisis was preceded by several years by attention to environmental damage. The public recognized a need to minimize the damage to our life-support system while continuing to develop fossil and nuclear energies. More energy resources were to be put into technology for protecting the environment. The Environmental Protection Agency (EPA) was set up, and laws were passed requiring more technological uses of extra energy to help the environment. But pollution from environmental technology may have been as great as the protection it afforded, and EPA seemed confused about whether it was protecting human beings from the environment or the environment from human beings. As we have already seen, when humanity and nature are adapted to low energy, recycled wastes are used as valuable sources of energy.

In many minds there has been a conflict between energy and economy on one hand and environment on the other. We have already seen that these are all part of the same system and that to survive, an economy must use the environment well—that is, must have a symbiotic relationship with it. To decide which measures help the total energy effect, including that affecting the ecosystems, we must calculate net energies.

Sulfur in Coal and Oil

When fuels are burned, waste acids are emitted. We call these *acid volatiles;* and in Chapter 8 we compared the emission of these volatiles to the action of volcanoes. Sulfuric acid is the most corrosive of them and is formed when sulfur dioxide gas from sulfur in the burning fuel interacts with water. This is the sulfuric acid most people may remember from their science classes; it eats holes in clothes, dissolves limestone, and burns the skin, leaving scars that heal slowly. Slightly acid rains have been found in all regions whose winds blow across industrial regions.

Low-sulfur coals and oils had been required in many localities in order to keep down the release of sulfuric acid and thus reduce corrosion and the threat to human health. But low-sulfur fuels are more scarce and cost more than high-sulfur fuels. Coal that used to run power plants has been displaced by oil partly because of the sulfur problem. As soon as the energy-inflation crunch developed, people began to reconsider the decision to require low-sulfur fuel.

The question to be answered is whether the overall increased energy costs to the community of low-sulfur fuels (measured in terms of energy, since the money paid for them can be considered as energy-buying ability) was greater than their negative effects on property, soils, and human health (also measured in energy terms). These calculations are being made. Certainly, in centers of dense population, particularly where limestone is used for buildings (as in Venice, Italy), use of low-sulfur fuel justifies its extra costs. In remote regions, and regions where soils have plenty of basic materials to neutralize the acids, high-sulfur fuel may be used with less effect.

Mining (See Figure 11-2)

Since coal is one of the large sources of net energy still available, another environmental issue has been whether to do massive strip mining. This is the process of removing overlying rock and soil to get at coal deposits. The coal layers, of different thicknesses, lie under many layers of other kinds of rock and sediments. The material above the coal is stripped off and piled aside. Then the coal is scooped out and shipped away. This process removes the land from the production processes of agriculture, vegetation, soil development, water management, and so on—the processes of the interaction of the earth's crust with the sun that we described in Chapters 7 and 8. If the land is left in great piles of broken rock and unorganized dirt, a very long time is required for the rebuilding of soil, groundwater flow, and vegetation, and the restoration of order and normal processes. However, some of the energy that is yielded can go back into regrading the land and doing as much as is possible to reestablish a workable system of vegetation, agriculture, soil development, etc. When this restoration is done immediately, fewer years are required for the water, winds, and plants to reestablish the fine structure of geological strata and soil.

If proposed strip mining is too deep, with too much overburden to be removed and too much restoration to be done, then it yields no net energy.

If the energy cost of getting energy, steel, and technology continues to rise, then machines will become too energy-rich to use to mine coal, and the older low-energy system of mining by hand will return.

When deposits are very deep, but concentrated, they may yield net energy if mined by tunneling. This method involves high energy costs which include safety measures and the costs of impaired health and accidents, in addition to the costs of tunnels and shafts.

Rarely are all the energy yields and costs estimated before choosing a method of, and a location for, mining. In addition to the costs we have mentioned, there are also slag piles, land slumping problems, and wastes of mine drainage. As prices encourage more coal mining again, the energy costs will seem higher than they did before, since we now have higher expectations about environment and health. Moreover, in earlier days of mining there were great virgin forests to supply timbers for the mine shafts and tunnels. These are mostly gone now, and the energy cost of the mine supports will no longer be subsidized by several hundred years of forest growth.

POPULATION

Human population has a generation time of about thirty years, so that changes in trends in reproduction take time to affect the total population. Many people suppose that growth of population controls growth of everything else and that population is one of the stimulators of economic development and thus of energy. During the early years of this century, nations encouraged growth in population as a means of accelerating their energy development. It is true that more people help a system survive in a period of rapid net growth (this is the principle of maximum growth; see Chapter 5). The growth-accelerative action

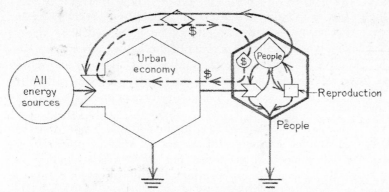

Figure 14-3 Relationship of population to energy support and urban economy. If energy sources are available, growth of the system can be stimulated either by more reproduction of people or by directly stimulating growth of the urban economy. Each accelerates the other but only if energy is available.

of population on itself was shown in Figure 5-3. Figure 14-3 shows the dependence of population on energy-processing technology, and vice versa, with both dependent on energy.

In the 1970s with energy scarce, adding population does not pump more energy but only divides the available energy among more people and causes unemployment. The "Zero population growth" movement started before the energy crisis and was at first individually oriented, to achieve the most energy per person. Some of its proponents hoped to help other parts of the world achieve as high a standard of living as the Western countries. Various mechanisms were developed for limiting population growth while still controlling disease; these have lowered the growth rate in countries with enough individual resources and education to follow the plans. Birth-control devices, abortion—and a general concern about the future food supply—have begun to reduce the birthrate. The leveling off of energy preceded the leveling off of population in some countries; in others, the reverse occurred. In the United States, many people thought economic growth was necessary to keep up with population, whereas it seems that the opposite is true: the population must slow down, since energy is declining.

CONFUSION IN RECOMMENDATIONS

Many different recommendations have been offered to solve the inflation-energy crisis. For example, it has been suggested that increased federal taxes be spent on federal programs. This suggestion has been partly motivated by the old idea of keeping an economy stimulated by accelerating money flows with more spending (see Chapter 4). Spending was a cure for a depression like the one in 1929, when there was plenty of energy in reserve. But in the current situation acceleration of spending makes more money flow for the same amount of energy—or less—and simply increases inflation.

Another suggestion is to balance the federal budget by decreasing federal spending, so that the money supply would not be increased. The size of decrease first suggested was less than 1 percent of the money circulating. This would hardly be effective where the rate of inflation is 12 percent and more.

Arguments have been made for cutting the military budget. Since the military establishment is not useful in maintaining foreign energy supplies at lower prices, the argument goes, it has little effect on world power. Less military reserve is needed for local defense against attack, since there are national stockpiles of bombs. Reducing the extent of military power to the actual borders of the nation has been proposed; but the idea has been opposed by people who remember the invitation to attack that isolationism created in the 1940s. However, that was a time of expanding energy and expanding military might all over the world. Now most of the world has declining net energy and thus a declining ability to affect others. Exceptions to this are the zones around the countries which still have rich oil, whose relative power position is increasing. But eventually their net energy will crest and then start to fall.

SUMMARY

In this chapter we have described the energy-inflation crisis as it developed in the 1970s, showing how declining net energy is changing the pattern of economic growth, eliminating large-scale capitalism, changing the balance of power and payments among nations, and shifting the world toward more decentralization. Disbelief in the energy crisis and misleading propositions for improving the economy were discussed. The energy bases for failures and false measures were shown. Many of the situations described in this chapter exist today or are still fresh in our memory.

We hope that the insights into the causes of economic change that come from energy diagrams will help the reader to understand what lies ahead, what is necessary for economic and cultural survival, and how the individual should plan his life.

Chapter 15

A Steady-State Economy

In this chapter we consider the properties of a steady state, which may be attainable now or may be something to look forward to with hope.

Because of the fears and misunderstandings that most people in the 1970s have when confronted with the idea of a decrease in growth, a leveling off of the economy, or a decline, there has been great difficulty in explaining the good qualities of steady states. The trouble may be that many people confuse the properties of organisms with those of the larger systems. A single organism has a period of growth followed eventually by death. But the larger systems are immortal if they have the energy to continually replace their parts and do not become too dense for easy substitutions of parts.

In Chapter 5 we described the steady state as a situation in which the inflows of energy balance the outflows. Examples from the physical world of the steady state in its simplest form were given in Chapter 5; one example is a bathtub in which inflow equals outflow. Examples from the ecological world were given in Chapter 7, where we described constant patterns of life in nature based on a balance of inflowing solar energy and outflows from work of maintaining life, replacing parts, and meeting various needs for survival. In Chapter 9 we examined how earlier human cultures lived in steady-state economies for long periods, in hunting and gathering societies and in farming cultures with agricultural villages. In preindustrial times, human beings lived either in a steady pattern or in a pattern of relatively small fluctuations and oscillations responding to minor variations of inflowing energy resources—sun, rain, winds, and so on.

CONDITIONS FOR A STEADY-STATE ECONOMY

Although the conditions under which steady energy sources produce steady states were considered in Chapter 5, we restate them here as they apply to our world of humanity and nature. Steady states develop when energy is supplied at a steady rate. Systems build the structures which maximize their use of energy and which they can also support on the energy available.

Under present energy conditions there are several ways in which human economy can approach a steady state. For a while, we could remain in a state like the present one, but without growth; later, we could become steady at a lower energy level. If the oil-producing nations supply fuels to the rest of the world at a fairly constant rate, and if nuclear energy continues to be only auxiliary, the rest of the world will move into a state that is fairly steady, for perhaps twenty-five years. This state will be temporary, since the ability of the fossil fuels to generate net energy will eventually decline. The economy would undergo a slow decline before becoming steady again at a lower level. When the rich geological storages of fossil fuels, metal ores, and nuclear fuels near the surface are mostly used up, the energy resources available as net energy to humanity and nature will be only those generated regularly by the sun. As we learned in Chapters 7 and 8, sunlight drives winds, ocean currents, waves, photosynthetic production, hydrological cycles, and cycles of sedimentary and volcanic materials. Certain materials, including a small amount of fossil fuels, will be available because of recycling by human beings or as a result of a slow regeneration of geological deposits by the mountain-building cycles. This may be as much as 10 percent of present usage.[1]

Figure 15-1 shows several alternatives for our future, including what will happen if we find any very large (unlimited) energy source (Figure 15-1b). Two transitions are shown from the present situation to the steady state. In one (Figure 15-1a), there is rapid, unlimited use of the remaining oil: this is "boom and bust." In the other, energy flow is managed by limitations on production and regulations of price at a high level, in order that the flow may continue as long as possible. This produces a partial steady state with a slow decline by one of the four pathways in Figure 15-1c, and later a more permanent steady state.

If the flow of energy were to stabilize at the present level, it would be possible to maintain much of the present pattern characterizing the United States, except the growth industries. The energy saved by no longer trying to grow should give a margin for transitional activities (such as helping people change occupations when necessary); and some energy would still be available for protection from other countries which had excess energy and were still undergoing growth.

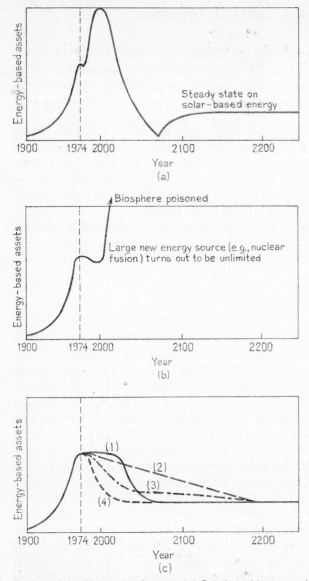

Figure 15-1 Alternative futures. (*a*) Continued boom and bust on present energy sources. (*b*) Unlimited growth. (*c*) Better patterns for humanity.

It is more likely, however, that some decline will take place before a steady state is reached, because people will not be convinced of the need to make changes until they are forced to change by energy shortages. The increasing energy cost of getting energy will gradually reduce the impact of the special energies and increase the role of lower-energy agriculture, forestry, fisheries, and other solar-based activities.

CHARACTERISTICS OF STEADY STATES IN ECOLOGICAL SYSTEMS

We obtain some insights into steady states from examining ecological systems in regions where climate and energy flows have apparently been continuous for millions of years. Rain forests, coral reefs, and the uniformly cold bottom of the sea (near freezing) are examples. In these situations there has been a very long time for the processes of organization and selection to produce a well-adapted system of species. We find great diversity; intimate, highly organized symbiotic relationships; organisms with complex behavior programs by which they serve each other; well-timed processing of mineral cycles that do not lose critical materials; and highly productive conversions of inflowing energy. There are moderate and steady storages of organic matter, information, and physical structure which are useful in making the systems go. Water running off from these systems tends to contain very little waste, since much of the materials of the system is recycled. Running on solar energy, these are low-energy, steady states.

CHARACTERISTICS OF STEADY STATES IN HUMAN COMMUNITIES

Low-Energy Steady State

The agricultural society in the Ganges valley in India before modern developments consisted of a fairly dense population tilling a variety of crop plots which were manipulated and irrigated to fit the seasons of flood and drought (this is a monsoon climate). See Figure 15-2. The famous sacred cows actually served special roles other than their religious one: controlling weeds, storing milk and protein, and providing animal power to pull plows for fast planting. The pattern was characterized by ritual; religion; markedly different customs for men, women, and children; considerable reuse of wastes; and considerable variety in crop plots. Population was in a steady state, controlled to some extent by disease and to some extent by social customs. Some goods were exported in trade. The system was lower in energy than fossil-fuel cultures, but the density of people actually completely supported on solar energy was about one to the acre. The system was low in density of energy per acre. Since there was external influence from invasions starting elsewhere, the system was not entirely without change; but there were generations that passed with little change. The culture was without much sense of progress, although there was a feeling that exchange was necessary and good.

Moderate-Energy Steady State

There have been a number of steady-state human cultures based on low (solar) energy; however, we do not have much experience with the steady state at higher energy. Many people advocate a steady economy; others fear being forced into one by a failing supply of energy. As is shown in Figure 15-1c, there is a good possibility that international handling of our current energy re-

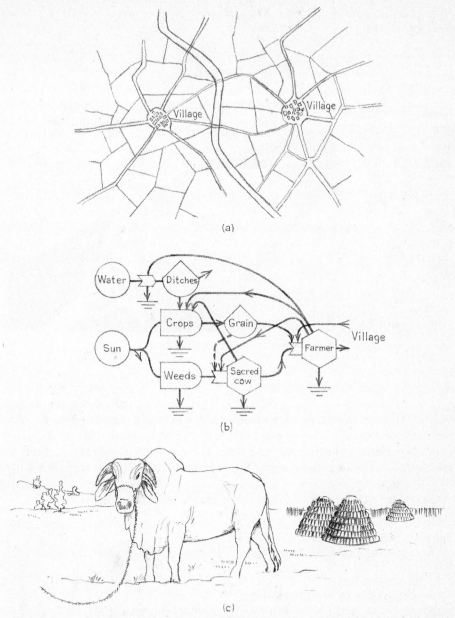

Figure 15-2   Agricultural energy patterns in India. (*a*) Layout of village, farm plots, paths, and canals. (*b*) Energy diagram. (*c*) Bull and stacks of manure drying for fuel.

sources may soon give us a short-term steady state or a gradual decline. Suppose that we do have a period of, say, twenty-five years with relatively constant energy made up of a combination of regulated release of fossil fuels, solar-based energy, and nuclear energy. In other words, let us suppose that our pat-

Gene Langley in The Christian Science Monitor © 1973 TCSPS.

tern is like Figure 15-1*c* instead of *a* or *b*. The following list gives some of the properties that we would find in a steady state which still used some fairly rich energy inflow. Some of these properties are due to leveling off and some to decline, especially to decline in net energy for new endeavors. (The reader is also referred to the section headed "For Whom the Bell Tolls," on pages 2–7.)

Characteristics of a moderate-energy steady-state economy

No net growth; little change after the transition to steady state; new items added only when old ones are removed.

Growth-stimulating industries are elminated, except for those at low levels.

Little advertising.

Less emphasis on transportation.

Loans, profit, and capital are minor; capitalism is essentially ended.

Money is no longer expected to earn money.

Stocks and banking have minor roles.

Governmental budgets are balanced.

Public works endeavor to maintain existing projects and buildings rather than initiate new ones.

Defense budgets are smaller, as are spheres of conflict.

Environmental technology that competes with ecosystems is discarded.

Wide diversity of specialization in retaining information that energy is inadequate to support everywhere.

Miniaturization of technology to use less energy.

Reduction in the number of competing businesses; increase in businesses that are in cooperative, symbiotic relationships.

Elimination of the feeling that progress is necessary.

More pride and pleasure in efficiency and performance without expansion or profit.

Decrease in public and private choices and experiments.

Shift of functions from large-scale to smaller-scale institutions.

Religion and personal ethics replace some law.

Social customs replace some governmental rules.

Smaller institutions are more dominant; there is more local individuality.

Smaller families; better use of both sexes in the work force; provision of some social needs by communal groups; decrease in need for reproduction.

Less travel and tourism; less luxurious entertainment; fewer large airplanes and large trains; less high-energy (rapid) travel.

Medical care more localized; less very expensive hospital treatment; more use of pharmaceuticals.

More reuse and recycling of all materials.

Urban construction will be replaced by separate and smaller houses, separately constructed and maintained.

Artificial, ornamental vegetation such as lawns will tend to be replaced by smaller-scale, more diversified food and fiber production.

Air conditioning will be replaced by architecture that fits human settlements into landscapes and vegetation so as to use natural means of cooling and geothermal heat.

Eutrophication as a problem will decline as nutrients from animal and human waste are sought for recycling in production of foods and fibers.

Universities will have fewer members and will diversify; they will hold as much knowledge as possible, do less new research, and teach more that helps individuals adapt to the new conditions. Universities may be critical during transition.

Alternative Sources of Energy.

Drawing by Niculae Asciu; © 1974 The New Yorker Magazine, Inc.

Farms use more land, less fuel, and more hand labor; small units will develop as use of machines decreases.

Large, regional power plants will be replaced with smaller, local ones.

Sewage plants will be replaced by smaller-scale, local recycling mechanisms which use more of natural ecosystems and agriculture.

Mental health will improve as each person is less affected by energy flows and change.

Properties of high concentrations of energy will decrease: crime, accidents, law enforcement, noise, central services, taxes.

Forestry will shift from fiber for plastic products and paper to lumber for building materials.

CARRYING CAPACITY

In Chapter 5 we discussed the steady level of water developed in a single tank supported by a steady energy supply (Figure 5-1). This was given as an example of a steady state based on a steady inflow of energy resources. The same idea has long been used in the management of wildlife, where it is called *carrying capacity*. Carrying capacity refers to the level of population of some species that a forest can support in the long run with its regular, renewable energy supplies from sun, rain, wind, earth movements, etc. Carrying capacity is the population level that can be supported continuously, harmoniously, and safely. The carrying capacity is *not* a population so dense that it jeopardizes the natural ecological mechanisms for maintaining the complex habitat.

In recent years the idea of a carrying capacity for humanity has been much discussed. The energy requirements of a human being were dramatized when men were put into space capsules where everything necessary had to be stored and transported at great expense. The earth is also a space capsule. On earth, life support for humanity is provided mainly by seas, forests, and grasslands rather than by technology. The air, earth, and water are still purified mainly by the regular turning of water cycles, geological cycles, and biological processes on land and in the sea.

The carrying capacity for humanity depends on more than supplies of food, water, and air; it includes all the functions of the biosphere—agriculture, industry, and waste processing—that help support human economies.

Efforts have been made to calculate how many people could be maintained on earth. Now we realize that the carrying capacity for humanity is that density which makes the system of humanity and nature competitive by processing the most energy. Systems that become overdeveloped do not maintain an effective economy. Crowded, overly dense systems get less help from the biosphere and must use their special resources unnecessarily to treat wastes, prevent disease, and protect themselves from confusion, crime, and disorder.

Unlike the resources for wildlife, the energy supplies for human economic development come only in part from the renewable inflows of sun, wind, rain, etc., within various regions. Human economic systems now import most of their energy, especially fossil fuels. The carrying capacity for humanity in an area is, then, the population that is sustained in the long run on the basis both of renewable resources and of outside energies that it can attract. Figure 15-3 divides the energy basis for humanity into two categories: (1) the renewable resources of an area and, interacting with these, (2) the special energies that must be attracted by the economic system as exchange or investment. Many people have erroneously believed that the world has a very high—almost an

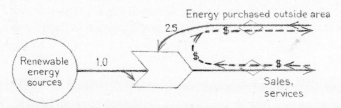

Figure 15-3 The energy basis in an area made up of renewable energy plus whatever the renewable energy can attract in economic exchange. Numbers give the ratio of renewable energy to purchased energy for the United States in fossil-fuel equivalents.

unlimited—carrying capacity, because they supposed that there was no limit to the fuels that could be attracted to an area.

Consideration of competition between areas gives us a formula for the carrying capacity of an area. Energy purchased depends on money earned from sales and services. The area that can best use renewable resources to aid production can charge lower prices, dominate the market, and purchase more energy from outside. An overdeveloped area has few renewable energy resources with which to help production costs. Prices are higher, income slows down, and energy flows from outside cannot be increased. By this reasoning, an area can increase economic development with purchases of energy from outside up to that point when the ratio of free renewable energy to purchased energy is that characteristic of its competitors. In the United States, this ratio is about 1.0 unit of renewable energy for every 2.5 units of purchased energy, where both are expressed in fossil-fuel equivalents. Areas with higher ratios of renewable energy to purchased energy tend to develop further economically; areas with lower ratios may tend to decline in economic development.

Since energies available for purchasing are diminishing, the proportion of renewable energies necessary to be competitive should increase in the future. By these calculations, a city such as Miami, Florida, may already be at its maximum size and energy level. New York may already be declining.

INHERITED STORAGES AND TRANSITION

The period of growth has left the world with enormous storages of high-energy assets such as highways, libraries, skyscrapers, power plants, large ships, and an enormous population. Eventually, declining energy will mean that not all these storages can be supported; but there will be a long period during which the storages already in existence will continue to be used. Gradually there will be selection for that which is to be maintained. We will have to find new uses for great urban centers, interstate highways, and so on. The challenge is to find new uses for the old structures so that they will have several functions and use less energy, and—even when they are not to be maintained—will have served some purpose. These large storages of assets are the means to prevent any sudden collapses, giving humanity time to make changes.

Everyone must think about the changes that are needed. Many laws may be inappropriate for the new conditions; perhaps there should be a general repeal of all statutes, restoring only those that are necessary in a simpler world. We should be able to wind some things down rather undergoing a collapse and then starting over.

STEADY STATE AS A HAPPY PLACE

Human beings were adapted to steady states earlier in our history and may be said to have had a good life of steady work and simple pleasures. There is a basis for the belief that the new steady state, or slowly declining levels of energy, will also give individuals a good life, with stability replacing the explosive growth and "future shock" of the recent past. If the best aspects of the growth period are selected and maintained with enough diversity, the less important and excessive aspects can be discarded. The kind of small-scale work, social interaction, and recreation that are characteristic of the steady state may be better for the individual.

To become adapted to the steady state, people will have to give up their restlessness and their insistence on the large, the new, and the different. But the young people who tried to form a low-energy subculture to avoid the excesses of the high-energy growth period will also have to change. More work will be expected from each individual in the low-energy society because there will be fewer machines. With less national and international organization of food, military activity, and information systems, local areas can make better, more individual adaptations to their own local energies. The richness of "one world" will be replaced by the richness of a great mosaic of diversified peoples and regions. As always, humanity is the most adaptable part of nature; and humanity is likely to fit itself to the energy flows of the earth in this, the newest system.

SUMMARY

In this chapter we have considered some possible future trends in our energy flows that involve little growth. Steady states or slow declines of energy levels will occur. Lower-energy patterns are likely to make better lives for individuals. We can expect a lessening of the tensions characteristic of high-energy growth—war, uneven distributions, disoriented individual relationships, disjointed production and consumption, and the dominance of machines. Indeed, it is fearsome to contemplate the possibility that we might return to fast growth if some large source of net energy should be developed. Wonderful challenges lie ahead for everyone in the transition to level or declining energy— opportunities to use our inherited storages well, to retain that which is best, and to regard these kinds of changes as progress of a better kind than explosive growth in which humanity may eventually destroy its basis of life support.

FOOTNOTES

1   Almost as much organic matter is buried in deposited sediments each day as is used as fossil fuel by human economies. Some small part of the deposited organic matter may recycle as newly regenerated fossil fuel. In other words, some part of the fuels we use can be regarded as renewable—it is in continuous formation and use.

Chapter 16

The Hopeful Future

In this chapter we summarize what we have said about energy, environment, and economics. We started with energy principles, considered the energy basis for humanity and the biosphere, and traced the changing pattern for humanity through the energy-inflation crisis. Humanity may be moving toward more regular energy flows, toward slowly declining energy flows, and eventually toward a steady state of unchanging energy flows.

Figure 16-1 summarizes the pattern of human economy in relation to the external energy flows of the biosphere. When we consider the energy basis of economics, we find that inflation is the warning and symptom of declining net energy and that the pattern of the future will be something like Figure 16-2. The wheel of economics must turn more slowly. We all must accept the facts of life—that is, the principles of energy—and prepare to live with less. The prospect of the steady state is, to us, challenging and enticing, and closer to individual human nature than the hectic period of growth which we are soon leaving.

THE ENERGY-MONEY WHEEL SLOWS DOWN

Figure 16-1 is a simplified view of the circle of work in the economy—human and natural—of the United States. Energy inflows from fossil fuels, nuclear fuels, the sun, rain, soils, etc., interact with the circulating goods, services, and materials of the human economy, the agricultural economy, and the economy of natural areas. All the energy contributes to the general turning of the energy wheel, which is performing the total work of humanity and nature. The energy, after its work is done, is degraded and can no longer be used for work; it disperses as degraded heat into the environment and ultimately into space. Although the energy enters the wheel of recycling materials, services, and information at particular places, its effect is dispersed throughout the system. Thus, the wheel accumulates the energies, distributing their work throughout the circulation, thereby making it hard to tell which of the energy inputs is contributing most. As the wheel turns, it serves to help pump in the energies at each point.

Circulating in an opposite direction, inside the energy wheel, is a flow of money that facilitates and lubricates the circle of energy. If one adds to the circulation of money, this may tend to lubricate and facilitate pumping of new energy by the energy wheel—but only if there are unused sources.

The true value of a circulating dollar is equal to the work that it causes in the energy wheel. This value can be determined by adding up the total Calories of energy (with all units expressed in fossil-fuel equivalents) flowing per year

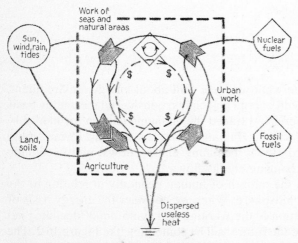

Figure 16-1 The cycle of money turning as a countercurrent driven by external energy sources. Energy that flows in spreads its work out over the whole circle. Payments of money go around the circle and cause the energy to distribute itself evenly around the circle. This diagram has the same symbols as those previously used, but here they are formed into a circle. (*Odum, 1973.*)

and dividing this number by the dollars flowing per year (the gross national product). In 1973 in the United States there were 25,000 Calories to each dollar (see Figure 4-2).

The energy inputs include the sun, winds, tides, rains, geological cycles of lands and soils, and many other elements that are not always considered in calculations. Money is exchanged for things that circulate from person to person. But it is not received by the energy sources; rather, it is passed to, say, the oil industry or agriculture for services of getting energy.

If a source of energy fails and the money in circulation remains constant, then the work each dollar represents drops by the same proportion as the decrease in the external energy sources. If some sources of energy become more dilute or more remote, the result will be the same as having less incoming energy.

Growth in energy—and thus increases in assets, goods, services, and so forth—has been possible in our recent history because new energy sources have been drawn in to support the work wheel at an ever-increasing rate. But now growth has subsided as the sources of energy have become limited, although growth can still continue where there are favorable arrangements to obtain rich oils.

It has been recognized that the net yield of fossil-fuel energy over and beyond that necessary to get the energy is decreasing; and this will eventually stop growth. Nations have begun to conserve supplies, and it has been possible for suppliers to hold back the richer sources and sell them for the same high prices as dilute inaccessible sources.

ENERGY DEVELOPMENT CRESTS

While economists are struggling to understand why the predicted return to growth is not forthcoming, and confused citizens still do not understand the energy basis of inflation, growth is stopping and the standard of living is declining. However, the course of the future seems clear to those who think in terms of net energy. Net energy flows in general are small, and those from nuclear energy still smaller. Figure 16-2 is a computer simulation of the rise and fall of assets as the energy reserves of rich fuels are used up. The turndown may be the trend for the whole world, but formerly rich countries are the ones to experience the turndown first. Growth continues where rich Middle Eastern oil is used.

The United States is leveling off; a turndown in total activity was seen in statistics on energy, money circulation (GNP), and productivity in 1974. However, many Americans are still talking about growth, normality, progress, and domination of international policies. A hazardous situation develops when a nation thinks it has more power than it really has. One theory for the decline of ancient Greece was that it tried to fight wars during a time when its energy, from exhausted soils and depleted forest resources, was declining. The

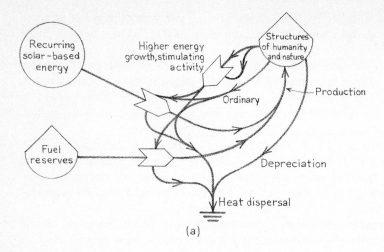

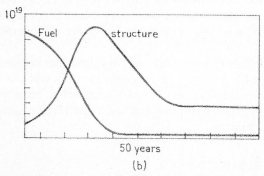

Figure 16-2 Simulation of the transition from a growth economy of fuel to a steady state based on the recurring solar-based energy flow. (*a*) Generalized world model of humanity and nature based on one-shot uses of fossil fuel and steady solar work. Here world fuel storage helps build a storage of structure of human buildings, information, population, and culture. (*b*) Graphs resulting from computer simulation of the model. (*Odum, 1973.*)

problem facing the United States is not whether or not to grow and find new energies, but how to decrease our use of energy and how to adapt to a situation in which less energy is available. Looking further ahead, we can predict a stable pattern potentially favorable to humanity; but this is not a growth economy.

The graph in Figure 16-2*b*[1], drawn by computer from the simple model in Figure 16-2*a*, resembles in many ways the computer-drawn graphs of Forrester and Meadows (Figure 16-3), which include many more details but have a similar shape. Both figures chart growth, leveling, and decline as rich resources are dispersed.

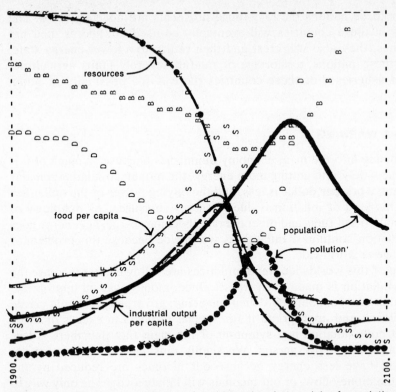

Figure 16-3 Computer simulation of world trends using models of population and resources by J. Forrester and his associates. "The 'standard' world model run assumes no major changes in the physical, economic, or social relationships that have historically governed the development of the world system. All variables plotted here follow historical values from 1900 to 1970. Food, industrial output, and population grow exponentially until the rapidly diminishing resource bases forces a slowdown in industrial growth. Because of natural delays in the system, both population and pollution continue to increase for some time after the peak of industrialization. Population growth is finally halted by a rise in the death rate due to decreased food and medical services." (*From* The Limits to Growth: A Report for the Club of Rome's Project on the Predicament of Mankind, *by Donnella H. Meadows, Dennis L. Meadows, Jørgen Randers, William W. Behrens III. A Potomac Associates book published by Universe Books, New York, 1972. Graphics by Potomac Associates.*)

REMAINING AREAS OF GROWTH

The crunch of leveling of growth in the West has been partly cushioned by purchases of military, nuclear, and technological know-how by the oil-supplying nations from the industrialized nations. This is very misleading, for these are one-shot purchases. When these exchanges of money are over, the decline in net energy and its consequent inflation could become more severe.

Growth has shifted to the nations supplying oil and critical raw materials.

But even for these nations the long-range prospects are not very good. The rich deposits in these countries will eventually be used up; and as their net energy declines, they also will crest and then return to a lower-energy state. In some of these nations, a shortage of rainfall severely limits agriculture, forestry, and fisheries. For these countries there is the possibility of boom and bust.

INFLATION—A WARNING

As energy flowing into the main economy diminishes because so much of it is fed back immediately into getting more energy, the money circulating remains the same. Thus work per dollar is less, and the buying power of the dollar has decreased. This kind of inflation is due to energy shortages. As rich flows of easily accessible fossil fuels and other energies become scarce, net energy goes down and inflation continues. Inflation is worldwide because this problem of energy shortage is worldwide.

If, on top of this worldwide trend, practices are followed that increase the money flow, inflation is made even worse. Once, money was an appropriate way to stimulate the economy by deficit financing; governments would create money with loans and appropriate it for special projects in excess of taxes taken in. But now, inflation is a symptom of the plain fact that there is less energy coming through per person. We simply must decrease what we do. Either we will have less energy because our incomes are reduced by the reduction of the money flow, or our incomes will inflate away. The only way to reduce this inflation is to reduce the money circulating as fast as the net energy decreases. To keep value of money constant, we have to plan a decreasing money supply. We must take some money out of circulation each year. This will help retain the value of saved money. But it must be realized that cutting back on our flows of money, salaries, and wages will force us to do less. If salary cuts are arranged for everyone, such reductions can be spread over the population instead of occurring as layoffs of a few. This will lessen unemployment.

ADAPTING TO DECLINE

We must realize the simple truth that energy per person must come down. The energy-rich activities at the end of the energy chain must be cut in proportion to the decline in sources at the start of the chain. Thus each individual's standard of living will come down. This will especially affect the rich, those who have made money from having money, those who are part of growth, and those involved with education, computer technology, and other high-quality activities. To put it bluntly, large-scale capitalism is ending, except in the energy-rich nations, which will have a few years more. Rather than letting inflation control the decline, we ourselves can control it by cutting salaries, decreasing the supply of money, and slowing down segments of the economy. Decreases in the money supply should match the inflation of the whole economy. This is now around 10 to 15 percent (100 billion dollars) for the United States. If we handle the reductions this way, cutting salaries at the same rate at which net energy declines, we can remain masters of what we do. Some money can continue to go to stimulate ideas about energy, in case they yield more energy than we now think; but the first priority for the next two decades is learning to live on the energies that we will have.

COST-BENEFIT ANALYSIS IN TERMS OF ENERGY

As energies decline, many hard choices will have to be made as to what is more energy-contributing and what is less energy-contributing. Decisions will be made by individuals about appliances, clothes, and housing. Decisions will be made at the national and state levels about public projects, tax policies that affect the money supply, and environmental protection as related to energy. These decisions concern energy-effectiveness and thus contributions to the vitality of the combined economy of humanity and nature; and they can be made by means of energy cost-benefit analyses. Most people are used to the idea of money cost-benefit analyses. We often compare money costs with money yields when we consider some action such as changing jobs or selecting a home. In an energy cost-benefit analysis we add up the positive effects of some activity on work done (in fossil fuel equivalents), including the work of nature and human work of man. Then we subtract the amounts of diminished work caused by the activity. This determines the net energy. From this we decide whether the activity is valuable.

If a good decision is the one that is most energy-effective in the long run, we should examine any proposed project for its effect on energy budget of the region, including both human beings and nature, since both are contributing to the vitality of the economy. Fossil fuel work equivalents can be estimated for the energy costs of technology and for its environmental effects.

For example, a power plant at Crystal River, Florida, was using estuarine water to cool its pipes, with some impact on the estuary. This impact was measured to be a reduction of half the biological productivity of 150 acres. A cooling tower was proposed. But when the energy cost of the cooling tower was calculated, it turned out to be greater than the energy impact of the estuarine cooling by a factor of 100 to 1. Wisdom in this case would select estuarine cooling, not the cooling tower.[2] Those in the environmental protection agencies who demand environmental technology instead of more symbiotic interactions with nature mean well but may be using funds for conservation poorly.

IMPACT STATEMENTS

Recently, many laws have been passed requiring those planning large developments to make a study of the impact that their proposals would have on the environment, on the economy, and on people. These statements have tended to be huge volumes of disjointed readings about the environment, each aspect considered separately. A better way of showing and understanding impact is now evolving, which considers the total system. Simplified diagrams and overall concepts such as energy can be used to bring everything into view and state everything in the same terms. Some of the diagrams of systems in this book are really impact statements. For example, Figure 11-2 shows the impact of strip mining and consumption of coal on the environment and the economy. In a particular situation, energy values are added to show what is important quantitatively. As energy and environmental values become more precious, justification of projects in terms of their impact will become even more important to ensure the thrifty use of all our energy flows.

DETERMINING COMPETITIVE PREFERENCE BY THE ENERGY-YIELD RATIO

Where our bought energies are to be used, we must decide not only if useful net energy will result from an activity, but whether this activity represents the best use of scarce net energies. If we must buy energy, we must sell something with the best possible energy support. In Chapter 6, we discussed the yield ratio: the ratio of energy yielded to energy fed back. It was suggested that we can compete only by keeping our sales, estimated in terms of energy cost, as high as or higher than competing parts of our economy. Decisions facing us during declining energy call for these calculations.

An example of such calculation is the question of offshore drilling in the stormy North Sea between Britain and Norway. Preliminary calculations have suggested that the energy involved in all the necessary construction and in the work of bringing oil ashore was as much as the oil yield (both stated in fossil-fuel equivalents). The energy-yield ratio was close to 1:1. In other words, this energy source is very energy-ineffective. If this is true, money would circulate with little useful work outside the oil industry. There would be a decline in the standard of living and the ability to compete; loans could not be paid back; and the plan would fail. The energy investment, then, would be better placed elsewhere.

POSSIBILITY OF A NEW GROWTH PERIOD

In looking to the future, we have considered the steady state and the state of slowly declining energy. We see little evidence that any existing or proposed energy sources will yield sufficient net energy to cause major growth after the rich fossil fuels are gone. We could be wrong about this. There is the possibility that large new energy sources will develop from nuclear fusion or from other theoretical concepts not now considered. In that case, we will have explosive further growth, environmental pollution on much larger scale, and a much greater likelihood of creating conditions unfavorable for human beings on the earth. The combination of machines, energy flows, and large biospheric disturbances of atmospheric, solar, and land systems would have results that are not pleasant to anticipate. It would become a genuine—and frightening— question whether humanity would stabilize its relationship to the planet and survive at higher levels of energy or would go beyond the survival threshold and be replaced by microbial and insect systems. The prospect of a steady state

"Don't tell me you want us to teach you to grow corn again."

Copyright © 1974 The Chicago Sun-Times. Reproduced by courtesy of Wil-Jo associates, Inc., and Bill Mauldin.

and low-energy life is much closer to what we know can make a happy existence for human beings and their planet.

If some new energy source were developed, the maximum-power principle would operate, and the high-powered system would displace lower-powered ones. If a major growth period should develop, it may not be favorable to humanity. One wonders how the specialists who are so infatuated with the idea of new sources of power can push so hard to use our remaining capital, storages, and net energy for this possibility.

THE WRONG WAY DOWN

We could make a mess of our transition if we fail to understand its nature. The terrible possibility before us is that there will be a continued insistence on growth, with our last energies, by economic advisors who do not understand. There would then be no reserves with which to make a change, maintain order, and cushion the impact on human life of a period when population must drop. At some point, great gaunt towers of nuclear power plants, oil wells, and urban cluster would stand empty in the wind for lack of enough technology to keep them running. A new cycle of dinosaurs would have gone its way.

There is a famous theory in paleoecology, called *orthogenesis*, which suggests that some of the great animals of the past were part of systems that were locked into evolutionary mechanisms by which the larger animals took over from smaller ones. The mechanisms became so fixed that they carried the trend toward largeness beyond the point of survival; thereupon, the species became extinct. Perhaps this is the main question of ecology, economics, and energy: Has the human system frozen its direction into the orthogenetic path?

Human beings will probably survive, because they reprogram readily. However, cultures which believe that only what seems good for man is good for the planet as a whole may perish.

Modern humanitarian customs regarding medical aid, famines, and epidemics are such that no country was allowed to develop major shortages of food and other critical energies; the other nations rushed in their reserves. This practice postponed realistic measures to limit populations. Now, when reserves are no longer there, the threat of starvation is great.

Chronic disease was evolved as a regulator of human life, being normally a device for infant mortality and merciful death in old age. It generally provided an impersonal and accurate test of vitality, adjusting the survival rate to energy resources. Even in the modern era of high-energy medical miracles, the energy for medicine is a function of the total energies of a nation. As energies fall, so will the energy available for medicine. The role of disease will again develop a larger place in the regulation of population. Chronic disease was, and is, a very energy-inexpensive regulator.

Epidemic disease is something else. Natural systems normally use the principle of diversity to eliminate epidemics. Epidemic disease is a device of nature for eliminating monoculture, which may be inherently unstable. Human beings are now getting special high yields from various monocultures, including their own high-density population, their paper pinetrees, and their miracle rice. These monocultures will survive only so long as human beings have special energies to protect them artificially from the disease which would restore the high-diversity system. Diversity is ultimately the more stable flow of energy.

People who are used to doing new things feel that they must approach the period of declining and leveling energy by doing something about it. This is proper if *fewer* things are done, but done *better*. There are many low-energy plans to be considered and many opportunities for smooth transitions. But applying technology (more energy) to the lowering of energy is an odd fallacy. In a growing economy with net energy there are plenty of resources with which to look to the future; in a declining economy, energies are obtained with great difficulty, and conscious efforts must be made to omit the less valuable activities before the energy to maintain the valuable ones is used up.

THE RIGHT WAY DOWN

It is exciting to watch as great changes in energy, economics, and environment unfold in detail day by day. Although we do not yet see much evidence of it, the energy theories suggest that there will be a large shift of population back to the land, with more feedback of human work to the solar energy chain. We will need plans for helping city people make the transition to farming.

Individuals will be needed more as machines do less. With less energy, there will have to be more intelligent application of our resources to make sensible transitions to the steady state. The best of our technology can be miniaturized and used with thrift. Nature will again help us, and more than ever before. We can achieve a better understanding for individuals to guide them in adapting to the increasing need for them. The road from energy crisis to steady state could be difficult but satisfying, with the peace of low energies at its end. Human beings will talk for many years of the flash fire of the twentieth century, its meaning, and its legacy for the continuing balance between humanity and nature.

SUMMARY

In this chapter we have summarized the main themes of the control of the economy by energy and the way we may turn away from growth. Our basis for discussion has been of energy resources and the fallacies in proposed alternatives to the steady state. We stated the danger of a new, unlimited energy source: it could extinguish humanity. We recommended ways of leading the world toward the steady state and discussed energy cost-benefit methods for making decisions about our scarce resources. Given time for a better adaptation, we see the future for humanity and nature as most hopeful.

FOOTNOTES

1 Available world fuel reserve was taken as 5×10^{19} Calories (net); and energy converted from solar input into human productive systems of growth and maintainence was 5×10^{16} Calories when structural assets were 10^{18} Calories. The peak of structural growth was variable over a fifty-year period, depending on amounts diverted into depreciation and waste.

2 Energies purchased from the general economy, in fossil-fuel equivalents, were 100 times the productivity of the estuary affected by the cooling. Compare this ratio, 100 to 1, with the average ratio for the United States (given in Figure 15-3), which is 2.5 to 1. Values much higher than the average are believed to be uneconomical and to represent a poor match of high-quality energy with lower-quality solar energy.

Glossary

Accelerate To cause to develop or progress more quickly.

Algae (algal cells) A group of plants variously one-celled, colonial, or filamentous, containing chlorophyll and other pigments and having no true root, stem, or leaf. Algae are found in water or damp places and include seaweeds, pond scum, etc.

Amplify To increase the output of one flow of energy by interaction with a second flow.

Aquatic Growing or living in or upon water.

Asset Anything that can depreciate and requires work to maintain.

Biomass The total mass (weight) of all organisms in an area.

Biosphere The zone of air, land, and water at the surface of the earth that is occupied by living organisms.

Carrying capacity Amount of animal life, human life, or industry that can be supported indefinitely on available resources.

Climax An ecosystem which is constant or repeating in pattern and replaces itself; an ecosystem at a steady state.

Consumer Organism, human being, or industry that maintains itself by transforming a high-quality energy source.

Culture Ways of life, language, social interaction, government, religion, etc., of a group of people.

Decline To lessen in force, value, function, etc.; used of storage, assets, energy.

Decomposers Consumers, especially microbial consumers, that change their organic food into mineral nutrients.

Delta   A deposit of sand or sediment, usually triangular, formed at the mouth of rivers.

Demand (economic)   The desire for a commodity together with the ability to pay for it; also, the amount of some commodity that people are ready and able to buy at a certain price.

Depreciation   A decrease in value of assets through deterioration.

Diversity   Variety; number of differences, as number of different species.

Ecological   Pertaining to a living environment.

Ecology   The study of environmental systems such as forests, lakes, seas, and urban areas, especially the living aspects.

Ecosystem   Ecological system. Examples: forest, coral reef, old field, pond, aquarium.

Energy   Quantity that accompanies all processes and is measured by the amount of heat it becomes; a quantity necessary for useful work.

Energy chain   A sequence of energy-transforming systems, each of which receives energy from the preceding system and supplies it to the next.

Energy conservation law   Energy flowing into a system is equal to energy stored in the system plus energy flowing out of it (all measured in heat equivalents).

Energy degradation law   Whether stored or being used, concentrations of energy spontaneously disperse, losing their potential for doing work.

Entropy   A measure of disorder, depreciation. (For an explanation of how entropy is measured, see footnote 2, Chapter 3.)

Environment   Surroundings; a zone of the biosphere occupied by ecological systems or human activities.

Estuary   An arm of the sea containing aquatic life, especially the wide mouth of a river, where the tide meets the current.

Ethic (as work ethic)   Idea or action which is considered right or moral.

Eutrophic   Term used to refer to a lake, pond, etc.: productive, rich in plant nutrients, minerals, and organisms, but often with variable conditions.

Facilitate   To make easy or easier.

Fallow land   Land left unfarmed for one or more growing seasons, to kill weeds, make the soil richer, etc.

Feedback   A flow from the products of an action back to interact with the action.

Fiber   Slender, threadlike tissues formed by plants or animals and used for paper, clothing, and building materials.

Fingerling   A small fish about the length of a finger, or a young fish up to the end of its first year.

Growth   Increase in size, weight, power, etc.

Heat   Energy in the form of motion of molecules; a form into which all other types of energy may be converted.

Heat engine   System drawing its energy from a concentration of heat (at a higher temperature than the surroundings).

Investment ratio   Purchased feedback energy divided by free natural energy where both are expressed in fossil-fuel equivalents.

Kelp bed   An ecological system dominated by any of various large, coarse, brown seaweeds belonging to the brown algae.

Kinetic energy   That energy of a body which is in its motion.

Maximum-power principle   Systems with more energy flow displace those with less flow.

Microorganism   An organism that can be seen only through a microscope.

Natural gas   Fuel gases obtained from decomposing organic matter in the earth.

Net energy   High-quality energy produced in a process in excess of high-quality energy used in the process.

Net production   More production than the consumption necessary for a process.

Nutrients   Mineral raw materials necessary for plant growth.

Oliogotrophic   Term used for a lake, pond, etc., low in plant nutrients, minerals, and organisms, low in productivity, with oxygen at all depths.

Organic   Derived from living organisms.

Oscillation   Regular fluctuation, variation, movement back and forth, as in ocean waves.

Peat   Partly decayed, wet, plant matter found in ancient bogs and swamps.

Perpendicular   At right angles to a given plane or line; exactly upright; straight up or down.

Photosynthesis   The production of organic substances and oxygen from carbon dioxide and water occurring in green plant cells supplied with enough light to allow chlorophyll to assist the transformation of the radiant energy into a chemical form.

Plankton   The usually microscopic animal and plant life found floating or drifting in the ocean or in bodies of freshwater, used as food by fish.

Potential energy   Energy in an inactive form that is the result of relative position or structure instead of motion, as in a coiled spring or stored chemicals.

Producer   Organism, human being, or industry that generates high-quality energy by transforming and combining low grades of sunlight and other energy sources and raw materials in excess of its own use.

Profit   Net increase in money; receipts in excess of expenditures.

Protein   The basic organic material of living matter, consisting of chains of amino acids; about 6 percent is nitrogen. Protein occurs in all animal and vegetable matter and is essential to the diet of animals.

Respiration   The processes by which a living organism takes in oxygen from air or water, distributes and utilizes it in oxidation of organic matter or food, and gives off products of oxidation, especially carbon dioxide and water.

Rural   Having to do with farming, agriculture, or country life; not urban.

Steady state   Pattern that is constant; it is based on a balance of inflows and outflows. Example: a river with an unchanging water level.

Subsidy   A payment from one unit of a system to another; aid, support.

Supplement   Something added, especially to make up for a lack or deficiency.

Succession   Sequence of stages and changes that occurs in an ecological system as it goes from some starting condition to a steady state.

Symbiosis   The intimate living together of two kinds of organisms where such association is of mutal advantage.

System   A combination of parts organized into a unified whole, usually processing a flow of energy.

Thermal waste   Waste heat from industrial plants that is discharged into the atmosphere or water.

Tundra   Vast ecological system of the treeless plains of the Arctic regions or high mountains above the timber line.

Urban   Pertaining to city areas.

Versatile   Adaptable to many uses or functions.

Weir   An obstruction placed in a stream, diverting the water through a prepared opening for measuring or controlling the rate of flow.

Work   Physical or mental effort exerted to do or make something; a force exerted for a
distance.

Yield ratio   Ratio of energy yielded to high-quality energy invested, where both are
expressed in fossil-fuel equivalents.

"Zero population growth" (ZPG)   A theory which advocates that there be no increase in
population, that each person only replace himself, and that birth control be
practiced in all countries.

Summary of Symbols

 Energy source from outside accompanied by causal forces. (See Chapter 1.)

 Heat sink, the draining out of degraded energy after its use in work. (See Chapter 1.)

 Energy storage tank delivers energy flow to pathways. (See Chapter 1.)

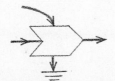 *Energy interaction*, where two kinds of energy are required to produce high-quality energy flow. (See Chapter 1.)

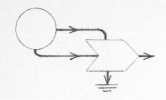

Super-accelerating flow with interaction of two flows from the same source cooperating. (See Chapter 5.)

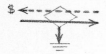

Energy-money transaction. (See Chapters 1 and 4.)

General-purpose box for any subunit needed; the box is labeled to indicate its use. (See Chapter 1.)

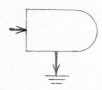

Producer unit converts and concentrates solar energy; it is self-maintaining. Details may be shown inside. (See Chapter 1.)

Consumer unit uses high-quality energy and is self-maintaining. Details may be shown inside. (See Chapter 3.)

Consumer unit requiring two kinds of energy interaction and input. (See Chapter 7.)

Mathematical Translations of Models in Chapter 5

Mathematical equations for the diagrams of Chapter 5 are given here for readers with backgrounds in physics, engineering, and computer simulation. The equations are for models 1, 2, 3, 5, and 6 (model 4 is not given; the student can work it out as an exercise).

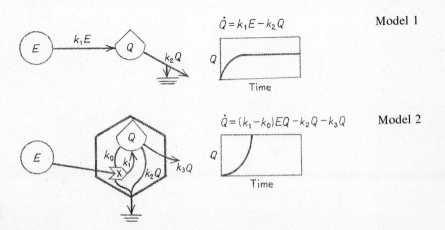

$$\dot{Q} = k_1 E - k_2 Q$$

Model 1

$$\dot{Q} = (k_1 - k_0)EQ - k_2 Q - k_3 Q$$

Model 2

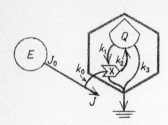

$$J = J_0 - k_0 JQ$$
$$\dot{Q} = (k_2 - k_1)JQ - k_3 Q$$

Model 3

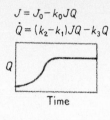

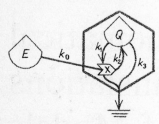

$$\dot{E} = -k_0 EQ$$
$$\dot{Q} = (k_2 - k_1)EQ - k_3 Q$$

Model 5

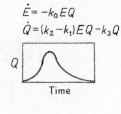

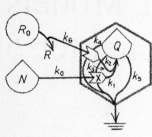

$$\dot{N} = -k_0 NQ$$
$$\dot{Q} = (k_1 - k_2)NQ + (k_3 - k_4)RQ - k_5 Q$$
$$R = R_0 - k_6 RQ$$

Model 6

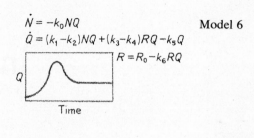

Exercises and Activities

Several concepts used here are from new research and are just now appearing for the first time in scientific reports and journals; therefore they may be new to most teachers. They include energy quality factors, fossil-fuel equivalents, the energy-yield ratio, and some of the overview models of humanity and nature. The language of energy diagrams is not new, but its use at this level is new.

The chapters are arranged to fit a one-semester course of fifteen weeks. If the book is used for a quarter course of 11 weeks, we suggest the use primarily of Chapters 1 to 6, which contain the basic energy principles; Chapter 7 or 8 for the environment; Chapters 10, 11, and 12 as background required to understand the energy-environmental-economic crisis; and Chapter 16 for the future.

Questions and activities for the chapters follow.

CHAPTER 1

1. Add up the total energy flowing into the farm (Figure 1-5) and that flowing out of the farm. How do the numbers compare? If outflows of energy are greater than inflows, what is the source of heat? Can this go on indefinitely? Is this a steady state?

2. Modify the diagram in Figure 1-5 so that fertilizer is added to the soil storage from outside sources at the rate it is used. Now answer question 1 for this new situation.

3. Use the symbols to diagram the flows and interactions of some other system, such as an automobile, a city, or town.

4   The systems diagrams are meant to help readers visualize and remember a system of parts. Reproduce the diagram of the farm, including the interaction of sources and the money exchange, without looking at the diagrams again. If you cannot do this, look at Figures 1-2 and 1-3 for a minute and then try again.

5   On the basis of your knowledge of farming, explain how the inflows accomplish the overall process of food production as diagrammed. Notice that the diagram does not show all the details that you may know about the farm, but it does include the overall necessary resources that may affect the overall process.

6   Draw the symbols for (a) an outside energy source; (b) an energy-flow pathway; (c) an energy storage place in the system; (d) an energy transformation process that involves more than one kind of inflowing energy; (e) a pathway of energy degradation that goes with any energy process.

7   What could happen to the amount of food produced if the price of human work goes up without any change in food prices? Hint: What change is there in energy inputs owing to the change in price?

8   Clip newspaper items that have to do with changes in our way of life that are related to energy.

9   Identify five questions about the current economic and energy situations that perplex you. Save these and see if you can answer them as the course proceeds.

10 Diagram a primitive subsistence farm which does not involve money (or machines).

CHAPTER 2

1   List the main kinds of energy flowing into and out of your home and your city.

2   Because rays of sunlight spread out from the sun, the concentration of sunlight is diluted with distance. What does a concave mirror do to the sunlight to affect its concentration? What kind of temperature results when one catches sunlight in such a mirror as compared with an ordinary flat surface? Try to make a piece of paper catch fire with a prism or curved glass.

3   Match up two of the types of energy from the following list for each of the energy transformation processes listed. For each process, which kind of energy is transformed into which other? Types of energy to chose from are: chemical potential energy, kinetic energy, potential energy of elevation against gravity, electrical current energy, light energy, energy in wave motions, heat.

Automobile operation
Solar water heater
Green leaf making food in sunlight
Sea breeze
A man skiing downslope
Water turbine generating household power
Windmill driving a water pump
Stove burning wood
Human being running after eating food
Steam engine running a train on coal fire that heats the boiler
Forest fire causing wind drafts

4   Name five processes which use the interaction of two or more of the following types of energy:

Kinetic energy
Heat distributed with differences of temperature

Potential energy of elevated water
Chemical potential energy in oil
Chemical potential energy in food
Energy contained in information flows
Energy in electrical currents
Energy in goods and services

5   Arrange the following types of energy in order of increasing quality: (a) food, (b) sunlight, (c) intelligent human work, and (d) dispersed heat.

6   Suppose fruits growing widely in a forest are gathered and concentrated by a boy and taken off by a wagon drawn by animals. The animals are fed some of the fruit as energy supply. Suppose that half of the fruit is eaten as energy cost in concentrating the fruit. If that fruit is the only cost, what change is made in the energy quality of the food by concentrating the fruit? What valuable energy contributions are required by the process that are not supplied by the fruit?

7   If 500 Calories of sunlight yield 1 Calorie of plant-food storage, what is the efficiency of production?

8   Suppose that a home receives in a day and uses 5,000 Calories of food, 50,000 Calories of electricity, and 150,000 Calories of oil. Suppose also that 100 Calories of the food are stored in new growth of a child. How much heat comes out of the house?

9   Diagram a fire from memory.

10 In Figure 2-5, which are higher-quality energy flows?

CHAPTER 3

1   State three laws of energy. Diagram and explain a process that includes all three.

2   People say, "The world is running down." What does this mean? Is it true for the earth? To what extent? Consider fuel reserves and the sun. What law does this refer to?

3   List the parts of these processes that are recycled and the parts that disperse into the heat sink:

You eat a steak.
A log is rotting.
A building is air-conditioned.
A geranium is flowering.

4   Pick a process. List the parts that you consider order and those that you consider disorder. Refer to Figure 3-1 if you need help.

5   Describe a system that illustrates the maximum-power principle.

6   Personal relationships are systems. What are some feedback energy flows in one of your own two-person systems? How do the feedback energies bring more energy into this system? Would you say that the relationship with "the most feedback to bring in more energy" has the best chance for survival? (Consider the relationship between a man and a woman, a parent and a child, or a boss and a worker.)

7   Think of all the energies that go into the relationships mentioned in question 6. Which are outside sources, and which are made within the systems?

8   Draw the symbol for a self-maintaining activity. Give two examples.

9   Give new examples of an unlimited energy source that can maintain constant force when used and a limited steadily flowing energy source. Make a diagram using one of them.

10 Attach the following labels to pathways on the general diagram for self-maintaining units. You may use more than one name for each pathway.

Depreciation
Production of higher-quality energy
Feedback
Entropy, a measure of disorder, increases
Pathway which puts a demand on the energy source by an interaction
Multiplier action
Interaction of two different energy qualities
Heat dispersed from work of production
Forces delivered from energy storage within the system
Pathway that uses resources to maximize power

CHAPTER 4

1 What is the ratio of money to fuel energy in the American economy taken as a whole? Suppose that a fisherman has to buy $1,000 worth of boats and nets each year. How much fuel energies have been spent in work on his behalf elsewhere in the human economy?

2 Do the circle of money and the circle of materials and minerals go in the same direction or in opposite directions?

3 What can one do to the supply of money to increase the rate of money circulating? What does this do to the total productive work when energy sources are unlimited? What does it do to production when energy sources are limited?

4 How would you increase energy flow by manipulating energy sources? How would you do it by manipulating the supply of money circulating?

5 What manipulations of money or of energy would cause the flows of work and money to level off (not grow)? To grow?

6 What happens to the yield of fish if the price of things that the fisherman needs increases? (Hint: Examine Figure 4-5 and see if one of the inputs to the fishing process is decreased or increased so that the fishing output might increase or decrease.)

7 Draw an energy diagram which has a farm, an industrial activity, and a population of people that are consumers and labor for the farm and industry. This diagram should have two energy sources: (a) the sun and related winds and rains; (b) the fuels that operate the industry. Show flows of goods and services from each of the three parts of the economy (farm, industry, and people) as facilitating the productive work of each of the others. This diagram should have each unit looped with the others. Finally, add money flow as a dashed line countercurrent to the energy flows.

8 In Figures 4-1 and 4-5, draw in the cycle of raw materials.

9 Using Figure 4-2, calculate the ratio of money flow to energy flow if there were no inflowing energy from fuel resources. How would this affect the standard of living?

10 If production stopped and assets declined 5 percent per year, what would happen to the buying power of money each year?

CHAPTER 5

1 Adjust inflows and outflows of water in a sink so as to reproduce the situations in Figure 5-2.

2 Given the model diagrams, visualize the nature of the growth curves that result.

3 Which of the systems drawn in this chapter is most like the present energy basis for the United States?

4 Can you identify which of the energy sources for man are unlimited and which are limited to a source-controlled flow? List them.

5 One important curve taught in ecology has the self-maintenance features of Figure 5-3b, with the outflow pathway a self-interaction flow. Can you draw the system? It is called the *logistic growth curve* and looks a little like Figure 5-4b. Can you think of a system that might be represented by this diagram?

6 One can set up a competitive-exclusion experiment by growing two species on the same rich food medium, keeping track of the number of each species on succeeding days. Your biology teacher may be able to help. Some possibilities are: two bacteria or fungi seeded into an agar nutrient plate; two species of grain beetle seeded into a bottle of flour; two species of grass planted in a well-fertilized flower pot and placed in the sun; two species of water fleas (or other plankton organisms) in bottles of water to which a pinch of yeast food is added daily (just enough to make the water slightly milky).

7 Prepare the electrical circuit shown in Figure A using wires, a solar photocell, an electrical storage tank (capacitor), a variable resistor, and a meter that reads electrical storage. This electrical circuit is an example of the model in Figure 5-2. See if you can reproduce the curves of growth on the meter by turning lights on and off.

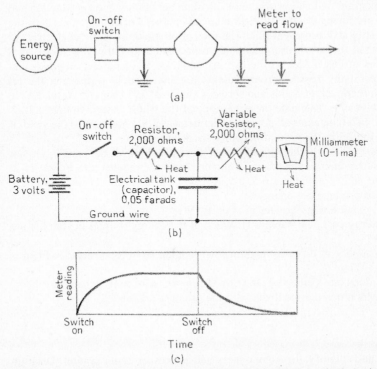

Figure A Suggested activity for an analogy with electricity: a single tank being charged by a source. (*a*) Energy diagram. (*b*) Electrical wiring (electrical symbols and terms are used). (*c*) Meter readings over time. Use battery or solar cell.

8 Find a new example for each curve. Draw and label a diagram for each.

9 Which diagrams end in a steady state? Why? (There are different reasons for different curves.)

10 Turn to the mathematical translations of the models in Chapter 5 on page 271. See if you can figure out how the diagrams are also ways of writing equations. (Hint: The rate of change of storage is $\dot{Q}$. This is equal to the sum of the inflow paths minus the outflow paths. Each pathway has a term in the equation.)

CHAPTER 6

1 What system are you familiar with that yields net energy? Diagram it.

2 In Florida, agriculture runs more on fossil fuel than it does on sun and other natural energies. What would have to be done to maintain the same yield if there were less fertilizer, fewer machines, fewer fancy varieties of crops, etc.?

3 Explain *fossil-fuel equivalents*. How are they useful in net-energy calculations?

4 For a farmer, what might be the advantages of growing a diversity of crops and the advantages of growing just one crop? How would you decide which to do?

5 If 1 barrel of oil contains 1.6 million Calories of fossil-fuel equivalents, and the average ratio of money to work in the economy is 25,000 Calories per $1, what would the cost of 1 barrel of oil be to yield no net energy? See Figure 6-8.

6 Figure 10-7 shows is an oil-company system like that in Figure 6-3, connected to an oil-fired power plant producing electricity. Identify three feedback loops. Locate a letter in the diagram that shows where to calculate net energy in oil production and net energy in producing electricity. How many fossil fuel work equivalents does the flow of electricity at Y represent if 3.6 Calories of oil make 1 Calorie of electricity? How many fossil fuel work equivalents are represented by the purchases at X_2 if 25,000 Calories of fossil fuel on the average generate $1 of purchased goods and services? How many fossil-fuel equivalents are involved in supporting the oil company with feedback of goods and services shown at X_1? Using Y, X_1, and X_2 in fossil-fuel equivalents, calculate the net energy of the whole system in Figure 10-7. Which flow is called an *external flow* (see Chapter 4)? What happens to the yield of electricity as the oil in the ground becomes harder to get?

7 Define *net energy*. Draw a diagram of a coal mine which yields net energy. Do not forget the "hidden energy subsidies." Then compare the results with Figure 11-2.

8 Give the yield ratio for Figures 6-5, 6-6, 6-7, 6-8, 11-2*b*, 11-5*c*, 11-6*b*, and 11-7*b*. Which of these produce net energy?

9 In Figure 7-9, is net energy greater at A or B? Why?

10 In terms of energy cost (see Table 6-1), which of the energy sources in your life are most valuable?

11 Is the energy quality of diversity high or low, judging from its position in Figures 6-12 and 7-11? Why?

12 In Figure 6-1, does one system (A, B, C, or D) require more energy than the others? Per Calorie, which has the feedback with the greatest influence?

CHAPTER 7

1 Visit a pond and identify the components and processes in its system. Diagram them. Do the same for a forest.

2   Examine aquaria and terraria such as those kept in many schoolrooms. The usual aquarium has very little light and very heavy food consumption, with outside food added. Which figure in Chapter 7 is appropriate to describe this?

3   Count 1,000 trees in a forest or shells on a beach. While counting, select one of each type found and make a note of it. Is the number of types (species) found large or small? (See the text for an explanation of *large* and *small* diversity in nature.)

4   Describe the variation in dissolved oxygen to be expected in the course of a day and night in a pond. Explain with a diagram (refer to flows in Figure 7-1).

5   Visit your local sewage plant. Note the means of decomposing organic matter to nutrients. Where does the outflow go? What plants use the nutrients in the outflow? Diagram the whole system. Which figure in Chapter 7 is most appropriate?

6   Which diagram best represents a lake receiving waste from paper making?

7   What happens if one nutrient is in very short supply? Explain by referring to Figure 7-1.

8   Diagram a food chain for a lake. For a forest. For human beings in a modern city.

9   Describe succession for your region on land. What is the succession in ecosystems in fresh water?

10 Suppose one introduced a high diversity of species from widely different locations into an enclosed field. What do you think would happen to the diversity?

11 Which is probably larger in the whole biosphere during our present period of large urban activity, photosynthesis or respiration?

12 Collect pictures from magazines for two different types of ecosystems and show how the diagrams relate to them.

13 In Chapter 5, find an example of succession leading to a decrease in storage.

CHAPTER 8

1   Explain the principle of a heat engine.

2   Show what the steam engine and the atmosphere circulation have in common. Can you use the same diagram for both systems?

3   Draw a sectional diagram through a continent and describe the sedimentary cycle and how it connects with the cycle of spreading on the sea floor.

4   Name three sources of heat to the earth, as in volcanic action.

5   Why are the deserts of the world located at 30 degrees latitude? What is it about the general circulation that causes them to have so little water?

6   Where does a hurricane get its energy? How does it use its stored energy to pump in more?

7   What do the cold outbursts of frigid air that pour across the United States in winter behind cold fronts accomplish in the general circulation?

8   Compare Figures 8-5 and 8-6. What are the similarities and differences between air flows and ocean flows?

9   The most common continental rock is light-colored granite which contains much silica. Examine a piece and compare its density with that of rocks of the ocean floor (basalt). What does this indicate about continents? What happens to granite and how does it get into the sedimentary cycle and form again?

10 Explain: sea-floor spreading, ocean-floor blocks, floating continents, and sulfur-dioxide and sulfuric-acid pollution.

11 Show how the main layers of rock of your state fit into the earth cycles.

12   What kind of circulation would exist without life?

13   Compare the energy chains in Chapter 8 with the food chains diagrammed in Chapter 7. What parts of the earth cycles have the highest-quality energy?

14   Considering the role of water and chemical energy in raising mountains, is the energy quality of elevated mountains high or low? See Figure 8-8 and examine the table of energy quality in Chapter 6. (Answer: The quality is very high; one calculation is 1,000 Calories FFEs per Calorie.)

CHAPTER 9

1   Why can you call hunting and gathering society a *low-energy society*?

2   What is the difference between animal consumption in a regular stable climate and animal consumption in a sharply varying seasonal climate? Why is this so? (Hint: Figure 9-3.)

3   What is the advantage of diversity in a forest? In a group of people?

4   Diversity uses extra energy. What does this energy go for? Why is it worthwhile for a system to support diversity? Use the coral reef for an example.

5   A popular idea is "Parkinson's law," which says that individuals in larger organizations are less efficient than those in smaller ones. Is this so? Why or why not? Give examples, using organizations which you are a part of.

6   When human beings were hunters and gatherers, they were a small part of the natural ecosystem. How was agricultural society different? How was the natural ecosystem changed?

7   What is "shifting agriculture"? Under what conditions is it necessary?

8   Figure 9-7 shows Indians regulating finfish and shellfish. What does this mean? How was it done?

9   It is stated that the population of preindustrial human beings was controlled (often indirectly) by disease and hunger. This is also true in plant and animal communities. How does it work?

10   How does simple war between groups relate to the energy of each group and to their territorial boundaries?

11   The American colonies are an example of a system being built on the stored energies of previous systems. What two systems did the colonists use to support their early agriculture? What stored energies did they get from each system?

12   Diagram one of the systems in this chapter. Label each part.

CHAPTER 10

1   Using Figure 10-2 as a model, make a diagram of your village, town, or city. Be specific as to energy sources, manufacturing processes, sales, etc.

2   Progress and growth are considered good. Why? List four values in our society that agree with this. Can you think of four that disagree?

3   Redraw and label the parts of Figure 10-5 from memory and explain how growth generates capital and how capital is reinvested to cause more energy input and more capital. What happens to capital if there are no additional sources to be tapped?

4   What is the use of advertising, in terms of energy?

5   Population seems to have exploded beyond the point supportable by available energy. What is the cause of this explosion? What would you propose as solutions to the problem?

6 Define *profit, capital, interest, loans, principal,* and *assets* in terms of money. What do these mean in terms of energy?

7 How does the heat engine figure in the industrial revolution?

8 How does electrical power figure in the industrial revolution? What are the advantages of electrical energy in feeding back to do work on the production? What is the energy cost of generating electrical energy? Draw a model of the energy flows and feedbacks of an electrical power plant.

9 What differences are there in the energy basis for a city in a solar-based agricultural society and a city predominately based on fossil fuel? Discuss shifts in the role of cities going from one energy basis to the other.

10 Disease had a regulatory role in an agrarian economy. When the fossil fuels permitted increased populations to be supported, what had to be developed also from fossil fuels to change the natural regulatory system?

11 Draw a box representing your state and then show the main energy and money sources that are the basis for the system. Include money only if it crosses the boundary of the box. Link money to energy, including high-quality goods and services. Include renewable resources.

12 Suppose the government could adjust the circulating money in Figure 10-4 to be in constant ratio each year to the energy that flowed through and did work. Would inflation be possible? What would happen to the buying power of the dollar? Some people have suggested that we call the dollar an *energy certificate.*

CHAPTER 11

1 Collect pictures from magazines that illustrate each of our energy sources. Explain the energy significance of each picture to its energy supply systems.

2 Try to list the energy sources discussed in this chapter in the order of their apparent net energy without subsidy from the main energy flows of fossil fuel that run our main economy.

3 Which are the hottest heat sources that are harnessed by operating heat engines? Can all the potential energy in the high temperature be used in electric generators and other machines? Why or why not?

4 Can we use all the fossil fuels in the earth, including the very dispersed ones deep in the earth? If we do, what must we use?

5 Which fossil fuel will run out first in the United States? What will be the consequences of this?

6 How are we indirectly using tides for human life by using products of nature?

7 Name three nuclear processes, each of which is the subject of major efforts to develop plants. Which one creates the most dangerous waste? Which one is not working yet on earth?

8 Our present nuclear electric power plants are supposed to have fuels for how long?

9 Name five problems that make nuclear power controversial.

10 List three fossil-fuel energies, five energy sources directly based on the sun, one source within the earth, and one source which used the atom. Which of these are good net-energy yielders now?

11 Choose one of the energy sources discussed in the chapter. Find at least two newspaper or magazine articles about it.

12 What is meant by the chemical energy in pure water?

13 Why do solar technology devices have a poor yield ratio? What are proven ways of supporting human beings on solar energy?

CHAPTER 12

1 Give an example of each kind of nation in Figure 12-4. Specify the symbols by labeling them. For example, if you make the United States the nation in category 1, you might put "coal" in the storage tank labeled "energy."

2 What is "balance of payments"? Using two countries as examples, state their balance in energy and their balance in money. (It would probably help to fill in Figure 12-7.)

3 Why does the United States now have a less favorable balance of payments than it used to?

4 Human beings are considered territorial animals. The amount of territory each person or nation can have is related to its energy. How has each human being usually chosen and defended his territory (property)? How have nations defended theirs (boundaries)? Try to think how territories might be based on the realities of energy without conflict or war.

5 Examine the map of photosynthetic productivity of the land (Figure 12-2) and locate areas of high production. Can you correlate these areas with the areas of high population of all kinds of life in the history of humanity before fossil fuels were used? Were these areas those in which humanity could compete well with other organisms?

6 Progress may require periods of expansion of energy to yield extra energy that can go into innovation, research, individual freedom to explore new ideas, etc. Before fossil fuels were used, where were the surges of net energy? Were those areas the same as the areas of high production (Figure 12-2)?

7 During the period of colonization by Spain, much gold was introduced into European markets. What was the effect on the energy-equivalent value of gold? When, later, fossil fuels were introduced, making the energy budgets of nations much larger (see Figure 9-10), what happened to the power of gold to buy energy?

8 The map of solar energy reaching the Earth (Figure 12-1) shows large differences in different areas of the globe. Why are the areas of highest solar energy not the places of greatest plant production? How does the energy from the sun, even in the deserts, contribute to the energy basis of the biosphere in places where it is not first received? (See Chapter 8.)

9 If nations that once had the most powerful energy budgets are unable to get much fossil fuel or other special energy in the future, how will their military power be affected? What energy storage left over from periods of richer energy will affect military power for a long time?

10 Suppose that a nation determines to isolate itself economically, using only its own energy sources, which are not as concentrated and do not produce as much net energy as some outside sources? What will happen to the ability of this nation to protect itself?

11 Other things being equal, suppose that two nations have the same energy supply, but that one limits its population and uses more machines while the other limits its machines and has more people. Which has the higher standard of living for those employed (energy per person)? Which has the greater problem with unemployment? What other differences are there? Can you suggest some nations that may be in these categories? With which economy does miracle rice fit? ("Miracle rice" is a new variety with high yields which needs extra fertilizer and care.) Which has more land in agriculture?

12 Where are marine fishery yields large? (See Figure 12-9.) Explain the role of water depth on fish production. Explain the role of upwelling water that brings nutrients. How are fossil fuels and fish harvests linked? (See Figure 4-5.) Which countries can fish more? Why?

CHAPTER 13

1 Diagram a system containing an individual. Label the parts.

2 Using population of an underdeveloped country as an example, show how the goal of survival of the system may conflict with the survival of the individual. How does this fit with your philosophy?

3 How does individual free will fit with the idea that each person's purpose is to add energy to his or her system?

4 What is your share of the energy flow in the United States, if you do an average amount of work?

5 Why are your income and expenditure an incomplete measure of your energy basis?

6 In a growing economy (growing energy base), how should you invest your savings? In an economy that is declining in energy base, how should you invest your savings? What about a steady-state economy?

7 In which economy—growing, declining, or leveled—is there more energy for art? In which economy is art in short supply? What are some ways that the energy flows of art may be feeding back, affecting the basic productive energy flows?

8 Diagram yourself in the system of occupations and life that you may be in when you leave school.

9 Show how the order-disorder balance is involved in the life of a housewife. (See the order-disorder model in Figure 3-1.)

10 Identify occupations that go with growth and may have to decrease in number in a situation of declining or leveling energy.

11 What opportunities for the individual are there on farms in periods of growth, leveling, and decline as the use of fossil fuel first increases and then decreases? What kinds of farms will there be, and how much labor relative to machines will there be? What kind of farm families or other kinds of groups might there be if the population is not increasing, if medical protection for children is good, and if individual human beings need a dozen meaningful personal relationships?

12 The graphs of energy distribution in Figure 13-6 are called *power spectra* in some sciences, where they refer to energy chains of waves in the sea or turbulent motion in the air. Explain.

13 Use the numbers in Figure 13-4 to explain why it has not been possible to support a human being in a space capsule on solar energy.

CHAPTER 14

1 Many people still think that science and technology can solve all problems. Why may new technology not be the answer to the energy crisis?

2 What is the basic cause of inflation? How does this relate to the energy crisis?

3 List some indications that the economies of the United States and Florida are no longer growing. What are some indications that some growth is continuing?

4 Discuss "Project Independence," the federal policy that the United States should become independent in its energy supplies as soon as possible.

5 Explain, in terms of energy, why the stock market falls and rises.

6 How does the decline in net energy affect you personally? Consider goods, money, activities, education, etc.

7 A shortage of energy has been causing unemployment. What are the reasons for thinking that the supply of jobs may eventually increase rather than decrease?

8 Evaluate these recommendations for solving the energy crisis: (a) Spend federal taxes on federal programs. (b) Balance the federal budget. (c) Cut the military budget. (d) Cut personal, individual spending. (e) Put much federal money into exploring new energy sources. (f) Get ready for a life less dependent on fuel energy.

9 Draw a money flow on Figure 13-5a. At which end of the chain is the money received highest for each Calorie FFE flowing?

10 Use Figure 13-5 to suggest why efforts toward industrialization in India could produce more unemployment than elsewhere?

CHAPTER 15

1 The steady state can be defined as a system in which energy inflows equal energy outflows. Give examples from the physical world, ecosystems, and early human societies. What are the inflows and outflows in each of your examples?

2 What do you think a present-day steady state would be like for you personally and for the United States? Do you think it will be a "happy place"?

3 Name five industries that are dependent on a growth economy. Name five that could exist in a no-growth economy.

4 Choose two of the characteristics in the list on pages 246–248. Explain what they mean and how they fit into the steady state. Give examples.

5 Plan your low-energy house. Consider who will live there, how big it will be, what construction materials will be used, what utilities will be needed, what activities you'll do there, etc. Dream a little.

6 What sort of retraining should be given to Americans, to prepare them for permanent low-energy living?

7 If the average ratio of purchased energy to natural energy is 2.5 to 1.0 (both expressed in fossil-fuel equivalents), what is likely to happen to an area where the ratio is 1 to 1? What about the world as a whole, with a ratio of 0.3 to 1.0?

8 If you were given the job of calculating the carrying capacity of your county, what would be your purpose, what kind of facts would you need, and what kind of conclusions would you come to?

9 What human cultures of the past may have had three or more generations of steady state? Why do these not stand out in historical records based on large monuments and innovations?

10 Calculate how many Calories of fossil-fuel equivalents can be attracted each year to a country which is receiving 1 billion Calories of fossil-fuel equivalents per year as sunshine. How much economic flow is this in dollars per year?

11 The suburban apartments and houses that surround our cities are also next to rural areas. Can you suggest how they may be used in a new era when more people work on farms?

CHAPTER 16

1   Here is one suggestion for stopping inflation. Calculate the ratio of the total amount of energy flowing in the economy to the amount of money. As the amount of energy decreases, the amount of money would be decreased, to keep the ratio the same. The dollar would be worth a set amount of energy. All energies would be included—natural, fuel, and artificial. How does this proposal strike you? In what ways could this be done on a national level? Consider the politics as well as the economics.

2   Imagine that you are running for state senator. You realize that growth in your state has stopped. You want to help the people and the industries of the state to adjust smoothly. What would be your campaign platform? How would use these ideas to get votes?

3   Imagine that you are given the job of deciding whether the state road department should spend money on widening an existing road or put in a new, parallel highway 20 miles east of it. Your supervisor has told you to do an energy cost-benefit study. What would you include in your study? (Hint: See page 259.)

4   We have stated that the development of fusion energy could be "fearsome." Explain why you agree or disagree.

5   Proposals have been made in the United Nations and by many religious and socially oriented groups to send food to countries which are undergoing famines. What are reasons for and against this idea? How do you feel about it?

6   This book has a definite point of view toward the earth and humanity. Explain what you think it is. What parts of it do you agree and disagree with? Explain.

7   Suppose that the price of rich oil from the Middle East remains low enough so that countries which use it exclusively can continue to grow. What kind of work would these countries have to produce to balance their payments? What kind of future could a country expect if it followed this plan over a period of twenty-five years? Will the United States go in this direction?

8   Could a rural country which uses its fossil fuels only as military weapons invade and conquer a country which has more energy but is undergoing the stress of changing from a growth economy to a steady-state economy? Suggest other situations where military invasions can be analyzed in terms of overall energy resources. What areas of the world might develop sharp differences in energy?

9   What may be expected to happen to general levels of air and water pollution with a decline in the level of energy?

10 If institutions based on growth, such as social security, should fail, how could necessary social services be provided in an economy based on lower energy? What modifications of our welfare system could provide a gradual transition without bankruptcy? What must happen to the tax rates?

11 As energy levels are lower, what do you think about costs of doctors, hospitals, and medicine? (We cannot pay all the costs if the energy is not available.)

12 Miniaturization is one way to keep technology with less energy. Give three examples.

13 Refer back to the Introduction, where various suggestions are made to people in different occupations ("For Whom the Bell Tolls," pages 2–7). Can you explain the reasoning behind each point?

References

Ballentine, T. 1975. A Net Energy Analysis of Northern Great Plains Surface Mined Coal in Midwestern Power Plants. M. E. thesis, Department of Environmental Engineering Sciences, University of Florida, Gainesville.

Gardner, G. 1975. Preliminary Net Energy Analysis of the Production of Oil Shale and the Potential of Oil Shale as an Energy Source. Research report in ENE671, H. T. Odum, instructor, Department of Environmental Engineering Sciences, University of Florida, Gainesville.

Hirst, E. 1974. *Direct and Indirect Energy Requirements for Automobiles.* Oak Ridge National Laboratory, ORNL-NSF-EP-64.

Kylstra, C., and Ki Han. 1975. *Energy Analysis of the U.S. Nuclear Power System,* in H. T. Odum and C. Kylstra (Eds.), Report to U.S. Energy Research and Development Administration. Contract A +-(40-1)-4398. Pp. 138–200.

Leith, H. 1973. "Primary Production: Terrestrial Ecosystems." *Human Ecology*, vol. 1, no. 4, pp. 303–332.

Lem, Pong. 1972. Energy Required to Develop Power in the United States. Ph.d. dissertation, Department of Environmental Engineering Sciences, University of Florida, Gainesville.

McCune, S. 1972. "The Population Explosion and Its Effect on the Environment." *Proceedings of 1971 DuPont Environmental Engineer Seminar.* Bull. series 137. Engineering and Industrial Experiment Station, University of Florida. Pp. 1–16.

Meadows, Conella H., Dennis L. Meadows, Jørgen Randers, and William W. Behrens III. 1972. *The Limits to Growth: A Report for the Club of Rome's Project on the Predicament of Mankind.* A Potomac Associates book published by University Books, New York. Graphics by Potomac Associates.

Northern States Power Company. 1972–1973 (Winter). *Dimensions*, vol. 2, no. 1.

Odum, H. T. 1971. *Environment, Power, and Society.* John Wiley, New York.

Odum, H. T. 1973a. "Energy, Ecology, and Economics." *Ambio*, vol. 2, pp. 220–227.

Odum, H. T. 1973b. "Terminating Fallacies in National Policy on Energy, Economics, and Environment." In A. B. Schmalz (Ed.), *Energy, Today's Choices, Tomorrow's Opportunities.* World Future Society, 4916 St. Elmo Ave., Bethesda, Md. Pp. 15–19.

Odum, H. T. 1974. "Energy Circuit Models and Temperature." In J. W. Gibbons and R. R. Sharitz (Eds.), *Thermal Ecology.* U.S. Atomic Energy Commission Division of Technical Information, Oak Ridge, Tennessee. Pp. 628–649.

Riedl, R., N. Huang, and R. Machan. 1972. "The Subtidal Pump: A Mechanism of Interstitial Water Exchange by Wave Action." *International Journal on Life in Oceans and Coastal Waters,* vol. 13, pp. 210–221.

Sellers, W. D. 1965. *Physical Climatology.* University of Chicago Press.

Science. 1971. Frontispiece, vol. 172, May 21.

United States Congress, Committee on Merchant Marine and Fisheries of House of Representatives, 93d Congress. 1973. *Growth and Its Implications for the Future.* Part 2. Hearing Appendix, Subcommittee on Fisheries and Wildlife Conservation and the Environment. Serial no. 93-27.

Young, D., and H. T. Odum. 1974. *Energy Evaluation of Alternatives for Management of the Atchauxulaya Basin.* Report to Bureau of Sports Fisheries. Department of the Interior from Center for Wetlands, University of Florida.

Zuchetto, J., and S. Brown. 1975. "Evaluation of Alternative Solar Water Heaters," in H. T. Odum (Ed.), *Energy Basis for Environment Power and Society.* Report to Energy Research and Development Administration, June 1975.

INDEX